TP
265
.F42

W9-DFQ-375

Third Edition

Principles of Fire Protection Chemistry and Physics

Raymond Friedman

JONES AND BARTLETT PUBLISHERS
Sudbury, Massachusetts
BOSTON   TORONTO   LONDON   SINGAPORE

LIBRARY
WAUKESHA COUNTY TECHNICAL COLLEGE
800 MAIN STREET
PEWAUKEE, WI 53072

WITHDRAWN

World Headquarters
Jones and Bartlett Publishers
40 Tall Pine Drive
Sudbury, MA 01776
978-443-5000
info@jbpub.com
www.jbpub.com

Jones and Bartlett Publishers
Canada
6339 Ormindale Way
Mississauga, ON L5V 1J2
CANADA

Jones and Bartlett Publishers
International
Barb House, Barb Mews
London W6 7PA
UK

Jones and Bartlett's books and products are available through most bookstores and online booksellers. To contact Jones and Bartlett Publishers directly, call 800-832-0034, fax 978-443-8000, or visit our website at www.jbpub.com.

Substantial discounts on bulk quantities of Jones and Bartlett's publications are available to corporations, professional associations, and other qualified organizations. For details and specific discount information, contact the special sales department at Jones and Bartlett via the above contact information or send an email to specialsales@jbpub.com.

Copyright © 2009 by Jones and Bartlett Publishers, LLC.

All rights reserved. No part of the material protected by this copyright may be reproduced or utilized in any form, electronic or mechanical, including photocopying, recording, or by any information storage and retrieval system, without written permission from the copyright owner.

ISBN-13: 978-0-7637-6070-0

Publication of this work is for the purpose of circulating information and opinion among those concerned for fire and electrical safety and related subjects. While every effort has been made to achieve a work of high quality, neither the NFPA, the Publisher, nor the author guarantees the accuracy or completeness of or assumes any liability in connection with the information and opinions contained in this handbook. The NFPA, the Publisher, and the author shall in no event be liable for any personal injury, property, or other damages of any nature whatsoever, whether special, indirect, consequential, or compensatory, directly or indirectly resulting from the publication, use of, or reliance upon this handbook.

This work is published with the understanding that the NFPA, the Publisher, and the author of this work are supplying information and opinion but are not attempting to render engineering or other professional services. If such services are required, the assistance of an appropriate professional should be sought.

Production Credits
Chief Executive Officer: Clayton Jones
Chief Operating Officer: Don W. Jones, Jr.
President, Higher Education and Professional Publishing: Robert W. Holland, Jr.
V.P., Sales and Marketing: William J. Kane
V.P., Design and Production: Anne Spencer
V.P., Manufacturing and Inventory Control: Therese Connell
Publisher, Public Safety Group: Kimberly Brophy
Acquisitions Editor, Fire: William Larkin
Reprints Coordinator: Amy Browning
Manufacturing and Inventory Control Supervisor: Amy Bacus
Printing and Binding: R.R. Donnelley
Cover Printing: R.R. Donnelley

6048

Printed in the United States of America
12 11 10 09 08   10 9 8 7 6 5 4 3 2 1

Contents

Preface

Fire is basically a chemical reaction between combustibles and the oxygen of the air. The product of fire (smoke) consists of a complex mixture of chemicals that interact with humans trying to fight a fire or escape from a fire. Extinguishing agents and fire retardants are also chemicals. Clearly, an understanding of chemistry is a prerequisite to a thorough understanding of fire protection.

Certain branches of physics are also important in the understanding of fire. The rate of air mixing into flames, the buoyant rise of fire gases to the ceiling and the subsequent motion under the ceiling, and the escape of smoke from the fire compartment into connecting compartments are all key elements of the behavior of a fire. The rate at which heat is transferred from the flames to not yet ignited material or to humans trying to escape the fire is also clearly an important aspect. Finally, any attempt to understand the computer models now available for predicting fire behavior requires a knowledge of the underlying physics as well as the underlying chemistry.

Part I of this book is an elementary review of selected fundamentals of chemistry and physics that are most relevant to fire. Some readers who are already well trained in chemistry and physics will want to skip this part and proceed directly to Part II. Other readers might appreciate the opportunity to refresh their memories on basic aspects of chemistry and physics. Many terms in the text are defined in the glossary at the back of the book.

For those readers who have never been exposed to chemistry or physics instruction, it will be difficult to obtain complete mastery of all concepts in this book. Such readers would benefit from studying an introductory textbook on chemistry or physics, such as *Introduction to Chemistry*, 7th ed., by T. R. Dickson (J. Wiley, New York, 1995, 589 pp.), or *Fundamentals of Physics*, 5th ed., by D. Halliday, R. Resnick, and J. Walker (J. Wiley, New York, 1997, 984 pp.).

The bulk of this book describes the fire characteristics of materials (gases, liquids, and solids), the properties of combustion products (temperature, smoke, toxicity, and corrosivity), fire extinguishing agents and procedures, and movement of smoke. In addition, special fire situations (e.g., spontaneous ignition, exothermic materials, and fires in abnormal environments) are discussed. A chapter is included on computer modeling of fire. Additionally, most chapters close with a set of problems that offer a good self-review of the material just discussed. The answers to those problems appear at the back of the book.

The hardware and tactics of fire fighting are not treated in any detail, because principles of chemistry and physics are seldom involved. References on other sources of information on fire fighting are provided at the end of each chapter.

This book is the successor to *Principles of Fire Protection Chemistry*, 2nd ed., by the same author and publisher, and contains an updated version of the same material plus four new chapters covering the physics of fire.

Acknowledgment

The encouragement provided by my wife, Myra Friedman, was crucial.

Raymond Friedman
Palm Beach Gardens, FL
December 1998

PART I

Review of Fundamentals of Chemistry and Physics

SI System of Units

The international system, originally called the metric system, was introduced in France by Napoleon at the beginning of the nineteenth century. In the 1960s it was refined further and certain units, referred to as *SI units*, were agreed upon. *SI* comes from the French phrase *Système international d'unités*.

All industrialized countries, with the exception of the United States, have chosen SI units to express mass, length, time, force, energy, temperature, and other measures. Adoption of SI facilitates the following:

1. Communication
2. Exchange of manufactured products from one country to another
3. Computations, due to the use of factors of ten for each unit; instead of 12 inches to a foot and 5280 feet to a mile, SI uses 100 centimeters to a meter and 1000 meters to a kilometer

In the United States the primary users of SI units at this time are scientists; in the rest of the world, however, both scientists and ordinary citizens primarily use SI units or are changing over to their use. All engineering students in the United States are now trained in the use of SI units, and most advanced engineering textbooks use SI units primarily or entirely.

This text includes both SI and English units. In the more fundamental or scientific sections, only SI units are used. In the application sections, both SI and English units are provided in many cases. The U.S. reader should learn to convert back and forth between the two systems.

LENGTH, AREA, AND VOLUME UNITS

The basic SI unit of length is the *meter* (m). Originally, the meter was selected as $1/10,000,000$ of the distance from earth's equator to the North Pole. Table 1.1 shows various units related to the meter, with English equivalents.

TABLE 1.1   SI length units as related to meter, with English equivalents

SI unit	Abbreviation	Meter equivalent	English equivalent(s)
Kilometer	km	1000 m	0.621 mi; 3280 ft
Meter	m	1 m	39.37 in.; 3.28 ft
Decimeter	dm	0.1 m	3.94 in.
Centimeter	cm	0.01 m	0.394 in.
Millimeter	mm	0.001 m	0.0394 in.
Micrometer	μm	$m \times 10^{-6}$	0.0000394 in.

Area is expressed in square meters (m^2), square centimeters (cm^2), and so on. Land areas are expressed in hectares (ha); 1 hectare is 10,000 m^2. The English equivalent of the hectare is 2.47 acres.

Volume is expressed in cubic meters (m^3), cubic centimeters (cm^3), and so on. The liter (L) is commonly used as a unit of volume and is the same as 1 cubic decimeter (dm^3) or 1000 cubic centimeters (cm^3). A liter is equivalent to 0.264 U.S. gallons.

MASS AND DENSITY UNITS

The basic SI unit of mass is the *kilogram* (kg). The kilogram was selected because it is approximately the mass of 1 liter of water. The gram (g), also widely used, is $1/1000$ of a kilogram and is approximately the mass of 1 cubic centimeter of water. (Water expands or contracts slightly as its temperature changes.) *Density* (mass per unit volume) generally is expressed in grams per cubic centimeter (g/cm^3) or kilograms per cubic decimeter (kg/dm^3). The numerical value of density is the same in either of these units. The term *specific gravity* refers to the ratio of the density of a substance to that of liquid water.

Confusion often exists between mass and weight. *Weight* refers to the force acting on an object because of gravitational attraction and is a convenient way to measure mass on earth at sea level. However, if an object were on the moon, its weight would be only about one-sixth of its weight on earth; if the object were in an orbiting space station, it would be weightless; its mass would be the same in each case. This fact can be proved by an experiment measuring the inertia of an object (the force needed to accelerate it). The mass of an object is the sum of the masses of its constituent atoms and is invariant (except in a nuclear bomb explosion, when mass changes into energy). Table 1.2 shows the relationship of various units to the kilogram (with English equivalents).

TABLE 1.2   SI mass units, with English equivalents

SI unit	Abbreviation	Kilogram equivalent	English equivalent
Metric ton	t	10^3 kg	1.10 U.S. tons
Kilogram	kg	1 kg	2.20 lb
Gram	g	10^{-3} kg	0.0353 oz
Milligram	mg	10^{-6} kg	2.20×10^{-6} lb
Microgram	µg	10^{-9} kg	2.20×10^{-9} lb

TIME UNITS

Units for time are the same in the SI system and the English system. The basic unit is the second (s). Table 1.3 shows abbreviations for time units.

TABLE 1.3   Time units (SI and English)

Time unit	Abbreviation
Hour	h
Minute	min
Second	s
Millisecond	ms
Microsecond	µs

FORCE AND PRESSURE UNITS

The basic unit of force in the SI system is the *newton* (N). A newton is the force needed to accelerate a mass of 1 kg at the rate of 1 m/s/s. In the English system, 1 lb of force is the force that will accelerate 1 lb of mass at the rate of 32.2 ft/s/s. This definition was selected so that 1 lb of mass at sea level would feel a gravitational attraction of 1 lb of force.

From the relation between the pound of mass and the kilogram, and the relation between the foot and the meter, it is easy to show that a newton is equal to 0.224 lb of force. The gravitational force on 1 kg of mass at sea level is 9.81 N.

Pressure is force per unit area. The basic SI unit of pressure is the *pascal* (Pa), which is 1 N/m². One Pa is a very low pressure; therefore, a unit called *bar* also is used, which is defined as 100,000 Pa or 100 kilopascals (kPa). One bar is only 1.3 percent greater than normal atmospheric pressure at sea level; therefore, 1 bar is nearly equal to 1 atmosphere (atm).

Pressure in English units is often expressed in pounds per square inch (psi) or in inches of water. One bar equals 14.89 psi. One kPa equals 4.02 in. of water.

ENERGY UNITS

The basic SI unit of energy is the *joule* (J). A joule is the quantity of energy expended when a force of 1 N pushes something a distance of 1 m. Thermal energy as well as mechanical energy can be expressed in joules. One joule equals 0.239 calorie (cal), or 4.187 J equals 1 cal. (A calorie is the energy needed to heat 1 g of water 1°C; a dietitian's "calorie" is actually 1000 calories.)

Electrical energy can be expressed in joules; 1 J is 1 watt-second (Ws). One megajoule (mJ) (1,000,000 J) is 0.278 kilowatt-hour (kWh). For example, if a power of 1 kW is released in an electric iron for 0.278 h, then 1 mJ of thermal energy has been released.

In English units, energy is expressed in foot-pounds (ft-lb) or British thermal units (Btu). One ft-lb is equal to 1.355 J, and 1 Btu is equal to 1055 J or 252 cal.

POWER UNITS

Power is the rate at which energy is expended. In SI units, power is expressed in *watts* (W). One watt is 1 J/s. The kilowatt (1000 W) and the megawatt (mW) (1,000,000 W) are used frequently.

In English units, horsepower (hp) is still used. One horsepower equals 745 W. Also, note that 1 Btu/s is equal to 1.055 kW. (Roughly, 1 kW equals 1 Btu/s.)

TEMPERATURE UNITS

Two temperature scales are used in the SI system: the *Celsius scale* (°C) and the *Kelvin scale* (K). On the Celsius scale (at sea level), water freezes at 0°C and boils at 100°C. Negative temperatures are possible.

On the Kelvin scale, sometimes called the thermodynamic temperature scale, negative temperatures do not occur. The zero point on the Kelvin scale is called absolute zero and equals –273.15° on the Celsius scale. No temperature colder than this is possible.

Other features of the Kelvin scale indicate its basic nature:

1. The volume occupied by a gas is proportional to its temperature on the Kelvin scale, as long as its pressure is maintained constant (except at pressures far above normal atmospheric pressure).
2. The thermal radiation emitted by an opening in a hot furnace is proportional to the fourth power of the Kelvin temperature.
3. The velocity of sound through a gas is proportional to the square root of its Kelvin temperature.

Because of these and other scientific facts, it would be logical to use only the Kelvin scale for temperature. However, the Celsius scale (previously called the Centigrade scale) was used for more than a century before these facts were discovered, so the world continues to use both scales.

The Kelvin scale is expressed in "kelvins," not "degrees kelvin." To convert from K to °C, subtract 273. To convert from °C to K, add 273; that is,

$$°C + 273 = K$$

Therefore, for example, 20°C = 293 K, and –10°C = 263 K.

The English system uses the Fahrenheit scale (°F), where 0° is the temperature reached by mixing ice and salt, and 100°F is the body temperature of a chicken. On this scale (at sea level), water freezes at 32°F and boils at 212°F. Conversions from °F to °C are by the formula

$$\frac{°F - 32}{1.8} = °C$$

Conversions from °C to °F are by the formula

$$1.8(°C) + 32 = °F$$

For example, 86°F = (86 – 32)/1.8 = 30°C, and 25°C = (1.8 · 25) + 32 = + 77°F.

CONVERSION FACTORS

Chapter 1 provides formulae to convert °F, °C, and K temperature scales. Pocket calculators are available with built-in conversions for these and many other units. For the reader's convenience, Table 1.4 provides conversion factors for most of the quantities mentioned in this book.

TABLE 1.4 Conversion factors. To convert Column A to Column B, multiply Column A by Column C. To convert Column B to Column A, divide Column B by Column C.

Column A		Column B		Column C
U.S. unit	Abbreviation	SI unit	Abbreviation	Conversion factor
Inches	in.	Centimeters	cm	2.54
Feet	ft	Meters	m	0.305
Miles	mi	Kilometers	km	1.61
Miles per hour	mi/h	Meters per second	m/s	0.447
Square feet	ft^2	Square meters	m^2	0.0930
Acres	ac	Hectares	ha	0.405
Cubic feet	ft^3	Liters	L	28.3
Gallons (U.S.)	gal (U.S.)	Liters	L	3.785
Pounds (mass)	lb (mass)	Kilograms	kg	0.454
Tons	ton (U.S.)	Metric tons	ton (metric)	0.907
Pounds per cubic inch	$lb/in.^3$	Grams per cubic centimeter	g/cc *or* g/cm^3	27.7
Pounds per cubic foot	lb/ft^3	Grams per cubic centimeter	g/cc *or* g/cm^3	0.0160
Pounds (force)	lb (force)	Newtons	N	4.45
Pounds per square inch	psi	Kilopascals	kPa	6.90
Pounds per square inch	psi	Atmospheres	atm	0.0680
Atmospheres	atm	Bars	bar	1.013
Foot-pounds	ft-lb	Joules	J	1.356
British thermal units	Btu	Joules	J	1055.0
Calories	cal	Joules	J	4.187
Btu per second	Btu/s	Kilowatts	kW	1.055
Horsepower	hp	Kilowatts	kW	0.746
Btu/pound	Btu/lb	Joules per gram	J/g	2.33
Btu/pound-°F	Btu/lb-°F	Joules per gram-°C	J/g-°C	4.187
Btu/pound-°F	Btu/lb-°F	Calories per gram-°C	cal/g-°C	1.000
Btu/square foot-second	Btu/ft^2-s	Kilowatts per square meter	kW/m^2	11.35
Calories per square centimeter-second	cal/cm^2-s	Kilowatts per square meter	kW/m^2	41.87
Gallons (U.S.) per minute per square foot	gal/min-ft^2	Millimeters depth per minute	mm/min	40.7

Problems

1. Normal human body temperature is 98.6°F, and a 5°F increase represents a serious fever. Convert these values to °C. Convert to K.

2. A fire truck 12 m long is traveling at 90 km/h. Convert these values to English units.

3. A fire pump pressurizes water at 60 psi above atmospheric pressure and pumps it at the rate of 300 U.S. gal/min. Convert these values to SI units.

4. An electric motor self-heats 15°C above ambient temperature when operating steadily. If the ambient temperature is 70°F, what is the operating temperature of the motor in °F?

5. An automatic sprinkler provides water at a "density" of 0.3 gal/min-ft². Convert to SI units.

6. If the heat of combustion of benzene is 40 kJ/g, what is it in cal/g? In Btu/lb?

2

Chemical Elements and Compounds: Atoms and Molecules

WHAT IS AN ATOM?

There are 109 known chemical *elements*. Some of the more familiar elements are carbon, oxygen, hydrogen, helium, chlorine, iron, copper, lead, silver, and gold. Elements consist of *atoms*, which are tiny particles characteristic of the various elements.

Each atom consists of a relatively heavy *nucleus* having a positive electric charge, and a surrounding cloud of orbiting *electrons*, which are negatively charged and relatively light. More than 99.9 percent of an atom's mass is concentrated in the nucleus.

Each element consists of atoms unique to that element and different from the atoms of all other elements. Each element has been assigned an *atomic number.* The atomic number of hydrogen is 1; the hydrogen atom consists of a nucleus with an electric charge designated as +1, about which orbits an electron with a charge of –1. Helium is the second element, with an atomic number of 2; its atoms each have a nucleus with an electric charge of +2 (i.e., twice that of hydrogen). Two electrons orbit around each helium nucleus. The atoms of the third element, lithium, have a nuclear charge of +3 and three orbiting electrons per atom . . . and so forth, through the list of elements.

Earlier in this century this model would have been presented as a probable theory of how the elements differ from one another. Today, however, sophisticated scientific measurements confirm the correctness of the model.

Table 2.1 presents a partial list of the chemical elements, including most elements mentioned in this book. The properties of all 109 elements can be found in the *Handbook of Chemistry and Physics* [1], along with historical information on the discovery of each element.

One might infer, from the description presented so far, that a helium atom is two times as heavy as a hydrogen atom, a lithium atom three times as heavy as a hydrogen atom, and so on. Examination of the atomic weights shown in Table 2.1 makes it clear that this is not the case; for example, a lithium atom is about seven times as heavy as a hydrogen atom even though the electrical charge within its nucleus is only three times as large. This concept was difficult to explain until the neutron was discovered in 1930. The *neutron* is a particle with almost exactly the same mass as the nucleus of a hydrogen atom (which is called a *proton*), but the neutron is uncharged (electrically neutral) while the proton has a positive unit charge. The relative masses of the various kinds of atoms become understandable if each nucleus is considered to consist of a combination of protons and neutrons, which are bound together very tightly.

For example, if a helium nucleus consisted of two protons and two neutrons, it would have twice the electric charge and four times the mass of the hydrogen nucleus (a single proton). This is indeed the case. Likewise, a lithium nucleus consists of three protons and four neutrons and has seven times the mass of a hydrogen nucleus.

The atoms of the heaviest naturally occurring element, uranium, with atomic number 92, have 92 protons and about 146 neutrons in each nucleus and a mass 238 times that of a hydrogen nucleus. Notice the words "about 146 neutrons" in the previous sentence. Actually, a small percentage of uranium atoms have 92 protons and 143 neutrons in the nucleus. These atoms, which are slightly different forms of the same element, are called *isotopes*. The two isotopes of uranium are called U-238 and U-235, respectively.

The element chlorine is a mixture of two isotopes, Cl-35 and Cl-37. Each of these two isotopic forms has an electric charge of +17. The two isotopes contain 18 and 20 neutrons, respectively. The chlorine found on earth contains a larger proportion of Cl-35 than Cl-37, and the average atomic weight of chlorine is 35.45.

Actually, each element comprises several isotopes. However, a single isotope predominates for most elements, so that 99 percent or more of the atoms of that element belong to the dominant form. The atomic weights for each element shown in Table 2.1 represent the average of isotopes for that element as it occurs in nature relative to the C-12 isotope of carbon, which, by definition, is 12.

We must know atomic weights in order to calculate the combining proportions of reactants when chemical reactions, such as combustion, take place. It is important to memorize approximate atomic weights of certain commonly encountered elements.

TABLE 2.1 Abridged list of chemical elements

Atomic number (nuclear charge)	Name	Symbol	Atomic weight	Usual valence
1	Hydrogen	H	1.008	1
2	Helium	He	4.003	0
3	Lithium	Li	6.941	1
4	Beryllium	Be	9.012	2
5	Boron	B	10.81	3 or 5
6	Carbon	C	12.01	4
7	Nitrogen	N	14.01	3 or 5
8	Oxygen	O	16.00	2
9	Fluorine	F	19.00	1
10	Neon	Ne	20.18	0
11	Sodium	Na	22.99	1
12	Magnesium	Mg	24.30	2
13	Aluminum	Al	26.98	3
14	Silicon	Si	28.09	4
15	Phosphorus	P	30.97	3 or 5
16	Sulfur	S	32.06	2, 4, or 6
17	Chlorine	Cl	35.45	1
18	Argon	Ar	39.95	0
19	Potassium	K	39.10	1
20	Calcium	Ca	40.08	2
.				
.				
.				
35	Bromine	Br	79.90	1
.				
.				
.				
51	Antimony	Sb	121.8	3
.				
.				
.				
92	Uranium	U	238.0	3, 4, 5, or 6
.				
.				
.				
95	Americium	Am	243	3, 4, 5, or 6

The weights of the following elements should be memorized:

H = 1
C = 12
N = 14
O = 16

The column in Table 2.1 labeled Usual valence is explained later in this chapter.

HOW STABLE IS AN ATOM?

If an atom participates in an ordinary physical or chemical process, it can experience one of the following changes:

1. Lose a few of its electrons
2. Share a few of its electrons with a neighboring atom to form a chemical bond
3. Gain one or two extra electrons

Its nucleus, however, is completely unaffected. For example, if an atom is passed through a flame, an electric arc, or a laser beam, its nucleus will be unchanged. Atoms cannot be changed by fire, which produces temperatures of only a few thousand degrees Celsius at most.

However, if an atom is heated to a temperature of hundreds of millions of degrees, such as in a thermonuclear explosion or in the interior of the sun, then the nucleus can be changed. The original atom of element A can become an atom of element B, or perhaps it can split into two atoms of elements C and D. Another way to change a nucleus is to strike it with another nucleus that has been accelerated to nearly the velocity of light ($3 \cdot 10^8$ m/s) by the application of millions of volts of electricity. This book does not address the chemistry of nuclear transformations.

Finally, certain atoms, such as those of radium, are radioactive, which means that their nuclei are unstable. These atoms can spontaneously emit charged particles and become atoms of a different element.

WHAT ARE THE MASS AND SIZE OF AN ATOM?

Atoms are not infinitely small; they have finite, measurable size and mass.

Scientists who use refined equipment, such as mass spectrometers and X-ray diffraction cameras, are able to measure the mass and size of individual atoms. For example, the hydrogen atom has a mass of $3.3 \cdot 10^{-24}$ g and a diameter of $1.06 \cdot 10^{-8}$ cm. In proportion to their atomic weights, other types of atoms are more massive than hydrogen atoms.

These numbers can be understood through the following analogy: A very tiny droplet of water, produced by an atomizer, with a diameter of 1 μm and requiring a microscope to be seen clearly would have the same mass as 160,000,000,000 hydrogen atoms. If 10,000 hydrogen atoms could be lined up in a row, touching each other, then the row would be about 1 μm long. (A human hair is about 40 or 50 μm thick.)

WHAT IS A MOLECULE? WHAT IS A CHEMICAL COMPOUND?

Atoms are often bound to one another to form *molecules*. For example, if two hydrogen (H) atoms come together under the right conditions, they will form a hydrogen molecule, the formula of which is H_2. If a hydrogen molecule (H_2) encounters a hydrogen atom (H), one might expect that a molecule H_3 would form; this does not happen. Nor do entities such as H_4 or H_5 form. H_2 is the only stable molecular form of hydrogen. It is stable in the sense that it must be heated to 3000°C or more before it will dissociate (split apart) into two hydrogen atoms. However, the molecule H_2 can react chemically with other kinds of atoms or molecules at much lower temperatures, even at room temperature in some cases, to form different molecules.

Each element has its own "personality," in regard to forming molecules. The atoms of a number of other elements form stable diatomic (two-atom) molecules, including oxygen (forming O_2), nitrogen (forming N_2), and chlorine (forming Cl_2). However, not all elements behave this way. For example, neither helium atoms (He) nor argon atoms (Ar) combine to form He_2 or Ar_2. As another example of different behavior, three atoms of oxygen can combine to form ozone (O_3). However, ozone is not as stable as O_2 and will spontaneously decompose to O_2 within a few hours at room temperature.

A molecule, which is defined as a stable combination of atoms, can consist of either two or more atoms of the same element, as already discussed, or two or more atoms of different elements, in which case the molecule is a *chemical compound*.

For example, a hydrogen atom can combine with a chlorine atom to form the stable chemical compound HCl, or hydrogen chloride. (Hydrogen chloride will decompose back into its elements at temperatures above about 2500°C.) As another example, a molecule formed from hydrogen and oxygen is H_2O, or water. The H_2O molecule is quite stable and will decompose only above about 2700°C. Another molecule, H_2O_2, hydrogen peroxide, also can form from hydrogen and oxygen, but this molecule is much less stable than water and will gradually decompose at room temperature. Yet another molecule, OH, is not a chemical compound but a free radical because it is so extremely unstable. Free radicals are discussed in more detail later in this chapter.

The element carbon (C), a solid, can react with oxygen, a gas, to form either the gaseous chemical compound carbon monoxide (CO) or the gaseous chemical compound carbon dioxide (CO_2). Either of these compounds is stable by itself up to quite high temperatures, but either compound can react with other molecules to form new substances, even at lower temperatures.

This chapter has thus far concentrated mainly on gaseous molecules. However, liquids and solids can also consist of molecules. Water vapor (H_2O) can condense to a liquid by cooling, with the liquid consisting of H_2O molecules touching one another, whereas water vapor consists of individual H_2O molecules some distance apart, in a state of rapid motion, and frequently colliding with one another. The liquid form of water can freeze to ice, which consists of closely packed individual water molecules in a fixed orderly arrangement.

WHAT IS AN ION?

The atoms and molecules discussed so far are all electrically neutral; that is, the positive charges on the nuclei of the atoms are exactly balanced by the negative charges on the electrons associated with each atom or molecule. However, it is possible for an electron to become detached from an atom, converting the atom to a positive *ion*. Or, a free electron can attach itself to an atom, creating a negative ion. A molecule can also be converted to a positive or negative molecular ion by electron detachment or attachment.

If an ion has an excess or a deficiency of a single electron, it is referred to as *singly ionized*. However, sometimes the ion can have an excess or deficiency of two or more electrons, in which case it is *multiply ionized*.

Examples of ions are H^+, F^-, Mg^{++}, and O^{--}. Ions can exist in the gas phase, in liquid solutions, or in solid crystals, and they can react with molecules or other ions. To produce ions in the gas phase, extremely high temperatures, energetic radiation, or high-velocity particles are required. (All these conditions exist in an electric arc.) Once the source of ionization is removed, the ions and free electrons in a gas at atmospheric pressure will recombine within a fraction of a second to form neutral species. Ions can exist indefinitely in solution or in crystals.

STABLE AND UNSTABLE MOLECULES: FREE ATOMS AND FREE RADICALS

Consider a chemical compound of the elements carbon (C) and hydrogen (H), with the generalized formula C_xH_y. Such a compound is called a *hydrocarbon*. If $x = 1$ and $y = 4$, then the formula becomes CH_4, which is a stable molecule called methane. (Natural gas is mostly methane.) CH_4 is stable in the sense that it must be heated to at least 1000°C before it will decompose.

However, if $x = 1$ and $y = 1, 2,$ or 3, the molecules CH, CH_2, or CH_3 would exist. Each of these species can form momentarily, but none of them are stable unless frozen at a temperature such as that of liquid nitrogen (–190°C). At room temperature, molecules of any of these species will, within a fraction of a second, combine with themselves or with other available molecules to form new species.

If, in the formula C_xH_y, $x = 2$ and $y = 2, 4,$ or 6, this would represent the stable hydrocarbon molecules C_2H_2 (acetylene), C_2H_4 (ethylene), or C_2H_6 (ethane), respectively. None of these species will decompose at room temperature. On the other hand, the species C_2H, C_2H_3, and C_2H_5 are all extremely unstable, as are CH, CH_2, and CH_3.

Thus, certain combinations of atoms can occur to produce very stable molecules, while other combinations produce highly reactive species that can be preserved only by cryogenic (freezing) techniques. There also are intermediate cases where species such as ozone (O_3) or hydrogen peroxide (H_2O_2) can form with limited stability; these species can survive for hours or days at room temperature, and could last indefinitely if moderately refrigerated.

The unstable molecules mentioned previously, such as OH, CH, CH_2, and CH_3, are called *free radicals*. Certain kinds of atoms (e.g., hydrogen, oxygen, nitrogen, fluorine, chlorine) that normally form stable diatomic molecules (H_2, O_2, N_2, F_2, Cl_2, respectively) are called *free atoms* when they exist in unattached form. Free atoms are also unstable, in the sense that they rapidly recombine with each other or react with other available molecules.

Free atoms and free radicals are extremely important because they play a key role in high-temperature combustion reactions, mainly by participating in chain reactions, which are described on p. 42.

CHEMICAL BONDS AND VALENCE

The concepts of *chemical bonds* and *valence* provide a means of determining which molecules might form and, of these, which are likely to be stable and which are likely to be unstable.

Figure 2.1 shows six stable molecules, with chemical bonds represented by lines connecting the atoms. These are two-dimensional drawings of three-dimensional molecules, so they do not correctly show the relative positions of the atoms, which are known and could be represented with three-dimensional models. For example, the four hydrogen atoms in the methane molecule are actually arranged at the four corners of a tetrahedron, with the carbon atom at the center.

Figure 2.1 shows the concept of valence, which is the characteristic of an atom to form bonds with other atoms. One line emanates from each hydrogen atom, two lines emanate from each oxygen atom, and four lines emanate from each carbon atom. Therefore, hydrogen atoms have a valence of 1 (as do chlorine atoms), oxygen atoms have a valence of 2, and carbon atoms have a valence of 4. With these rules in mind, predictions can be made as to which combinations of atoms are stable.

Figure 2.2 shows the structures of four additional stable molecules. These each contain double bonds, as well as single bonds in some cases, while the previous group of molecules contained only single bonds. By introducing the concept of the double bond, we can explain the structure of certain stable molecules within the framework of the valence concept. According to the valence concept, each stable carbon atom has four bonds (either four single bonds, or two double bonds, or one double bond and two single bonds, or one triple bond and one single bond).

Figure 2.3 shows the structure of three free radicals, which are highly unstable molecules. These molecules do not satisfy the rules of valence, which specify that carbon should have four satisfied valences and oxygen should have two satisfied valences. Note that, for the *hydroxyl* radical (OH), one of the two oxygen valences is unsatisfied; therefore, it can be deduced

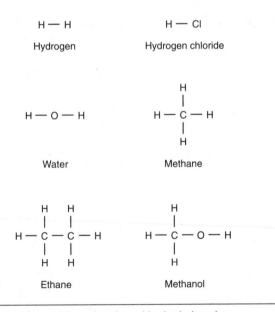

FIGURE 2.1 Six stable molecules with single bonds.

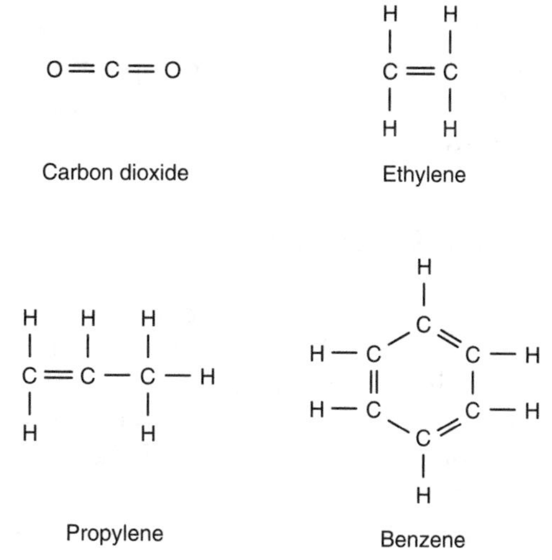

O=C=O

Carbon dioxide

Ethylene

Propylene

Benzene

FIGURE 2.2 Four stable molecules with double bonds.

that hydroxyl is highly reactive (and extremely unstable) and that two hydroxyls will combine rapidly to form H_2O_2 (hydrogen peroxide). Similarly, one of the four valences of carbon in the *methyl* radical is unsatisfied, and two methyl radicals will combine rapidly to form C_2H_6 (ethane). Or, a methyl radical will combine rapidly with a hydrogen atom, if available, to form a stable CH_4 (methane) molecule.

The valence principle presented thus far, while a very useful guide to molecular stability, nevertheless has exceptions. For example, the carbon monoxide molecule (CO) is quite stable up to several thousand degrees Celsius, and can be stored without change in a gas cylinder for years at room temperature. However, in carbon monoxide the valence of carbon is apparently 2 instead of 4 because the valence of oxygen is 2. (This molecule is the only important stable compound in which carbon has a valence other than 4.)

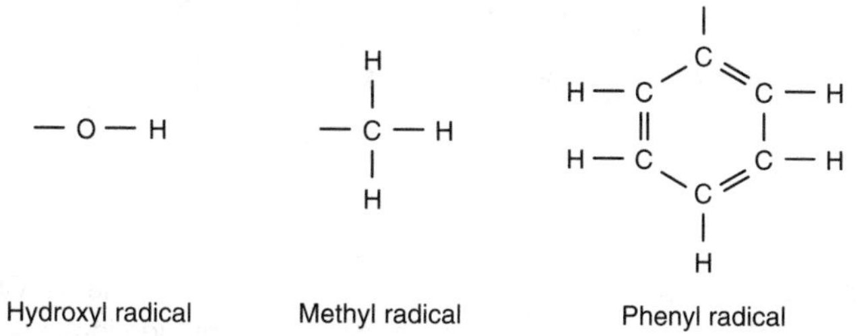

Hydroxyl radical

Methyl radical

Phenyl radical

FIGURE 2.3 Three free radicals.

Further, the valence of nitrogen in ammonia (NH_3) is 3, while its valence in nitric oxide (NO) is 2. Additional examples: The valence of copper can be either 1 (cuprous compounds) or 2 (cupric compounds), and the valence of iron can be either 2 (ferrous compounds) or 3 (ferric compounds). Table 2.1 shows the most commonly encountered valences of some elements.

CRYSTALS AND SOLUTIONS

Many solids exist in the form of crystals. A *crystal* is a geometrically regular array of atoms, molecules, or ions. In diamond, the elements of the crystal are individual carbon atoms. In ice, the crystalline elements are H_2O molecules. In sodium chloride (NaCl, common salt), the salt crystal consists of a cubic lattice with alternate positions occupied by positively charged sodium ions (Na^+) and negatively charged chloride ions (Cl^-).

Some crystals, such as sodium chloride, are held together by strong electrostatic forces. *(See Figure 2.4.)* Other forces are quite different and much weaker, such as those that hold together the molecular ice crystal; this is evident from the low melting point of ice (0°C) relative to the high melting point of sodium chloride (801°C).

If a sodium chloride (NaCl) crystal dissolves in water, no sodium chloride molecules will be found in the solution; instead there will be an equal number of positive sodium ions (Na^+) and negative chloride ions (Cl^-) moving freely in the solution. Each ion is bound to a cluster of surrounding water molecules, and these molecules prevent the positive and negative ions from recombining.

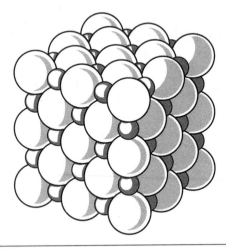

FIGURE 2.4 The atomic arrangement of sodium chloride (NaCl), a cubic crystal. The diameters of the sodium and chloride ions are $1.90 \cdot 10^{-8}$ cm and $3.62 \cdot 10^{-8}$ cm, respectively. [2]

A glucose (sugar) molecule $C_6H_{12}O_6$, however, dissolved in water, produces individual, neutral glucose molecules that move freely in the solution.

ISOMERS

The molecular formula of butane is C_4H_{10}. Figure 2.5(a) shows two different ways in which the 4 carbon atoms and 10 hydrogen atoms can be bonded together, while satisfying the rule that carbon has 4 valences and hydrogen has 1 valence.

These two forms of the butane molecule are called butane *isomers*. The configuration with four carbon atoms in a row is termed *normal butane* (*n*-butane), while the other configuration is termed *isobutane*. The physical and chemical properties of these two isomers are somewhat similar, but they are not identical. For example, *n*-butane boils at –1°C, whereas isobutane boils at –10°C. The spontaneous ignition temperature of *n*-butane in air has been reported as 408°C, whereas that of isobutane has been reported as 462°C. [3]

The second pair of isomers, shown in Figure 2.5(b), is dimethyl ether (CH_3OCH_3) and ethanol or ethyl alcohol (CH_3CH_2OH). Each of these molecules has the formula C_2H_6O, but the structure of each molecule is unique. In dimethyl ether the oxygen atom is bonded between two carbon atoms, and in ethanol the oxygen atom is bonded between a carbon and a hydrogen atom. The boiling points of this pair are –24°C and +78°C, respectively. The chemical reactivities of the two molecules are also very different. For example, the spontaneous ignition temperatures in air have been reported as 350°C and 558°C, respectively.

The third pair of isomers, shown in Figure 2.5(c), is cyclopropane and propylene, each with the formula C_3H_6 but with different molecular architecture. In liquid form, cyclopropane and propylene have densities of $0.72 g/cm^3$ and $0.61 g/cm^3$, respectively.

The fourth isomer example, shown in Figure 2.5(d), consists of the three isomers of dichlorobenzene ($C_6H_4Cl_2$), with the chlorine atoms in different relative positions on the benzene ring. The three isomers are as follows:

Isomer	Melting Point (°C)
Orthodichlorobenzene (*o*-dichlorobenzene)	–18
Metadichlorobenzene (*m*-dichlorobenzene)	–25
Paradichlorobenzene (*p*-dichlorobenzene)	+53

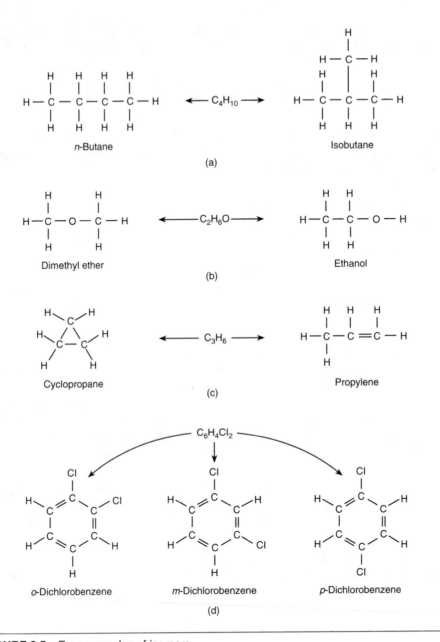

FIGURE 2.5 Four examples of isomers.

These isomers differ dramatically in that *p*-dichlorobenzene ("moth" crystals) is a solid at room temperature while the other two isomers are liquids at even well below room temperature. The difference in melting point between the ortho and para forms is 71°C (128°F).

From the viewpoint of fire science, the greatest concern is with the heat of combustion of these various isomers. Generally, there are only moderate differences. For example, the heat of combustion of dimethyl ether (CH_3OCH_3) is 3 percent greater than that of ethanol (CH_3CH_2OH). The heat of combustion of cyclopropane is 2 percent greater than that of propylene. Note, however, the spontaneous ignition temperatures of isomers can be significantly different.

Problems

1. How many different chemical elements are there? Are all the atoms of a given element identical? Explain.

2. Are all chemical compounds molecules? Are all molecules chemical compounds? Explain.

3. What is the relation between the valence and the stability of a molecule?

References

1. *Handbook of Chemistry and Physics*, 78th ed., CRC Press, Boca Raton, FL, 1997.

2. Pauling, L., *The Nature of the Chemical Bond and the Structure of Molecules and Crystals: An Introduction to Modern Structural Chemistry*, 3rd ed., Cornell University Press, Ithaca, NY, 1960, p. 7.

3. Mullins, B. P., *Spontaneous Ignition of Liquid Fuels*, Butterworths, London, 1955, p.117.

CHAPTER 3

Physical and Chemical Change

STATES OF MATTER

Three basic classifications of matter are found in the material world:

1. Gas (or vapor)
2. Liquid
3. Solid

These terms are part of the basic vocabulary of technology and science and usually can be identified by sight, smell, or touch. Steam and air are examples of gases. Water and mercury are liquids. Ice and iron are solid substances. Steam is a gaseous vapor of water, and air is the colorless, odorless gas in which we live and breathe. Vapors usually cannot be seen except in rare cases, such as the brown vapors of the element bromine or the greenish color of chlorine gas. However, the effects that are caused by gases when they blow dust and smoke around, or when they make rubber balloons expand, can be observed.

Properties of Gases

Gases consist of individual atoms or molecules moving at high velocities (approximately at the speed of sound, which is about 335 m/s in air). At atmospheric pressure and room temperature, the atoms or molecules themselves occupy only about 0.1 percent of the space, and the remaining 99.9 percent of the space is empty. Each atom or molecule is colliding with others and changing direction about 10^9 times per second.

If the temperature of a gas is increased at constant pressure, the average velocity of the atoms or molecules increases and the gas expands. If the pressure on the gas is increased by compression at constant temperature, then the molecular velocities do not change, but the density increases.

The density ρ of a gas is directly proportional to its pressure P, and inversely proportional to its absolute temperature T. The density also is proportional to the atomic or molecular weight M of the gas at a given temperature and pressure. These interrelationships are described by the *perfect gas law*, which is valid except at pressures far above normal atmospheric pressure:

$$\rho = \frac{MP}{RT}$$

where:

ρ = density (g/L)

M = atomic or molecular weight of the gas

P = absolute pressure (atm)

R = gas constant (0.08205)

T = absolute temperature (K = °C + 273)

Therefore, for oxygen gas at 20°C and 1 atm, the molecular weight M equals 32 [because the atomic weight of oxygen is 16, and the oxygen molecule (O_2) is diatomic], the pressure P equals 1 atm, and the temperature T equals 293 K (20°C). Then, the gas density equals

$$\frac{(32)(1)}{(0.08205)(293)} = 1.33 \text{ g/L}$$

Nitrogen has (N_2) has a molecular weight of 28 (2 · 14); therefore at 1 atm and 20° C, nitrogen has a density 28/32 times that of oxygen.

It is important to note that, at a given temperature and pressure, 1 L of oxygen contains the same number of molecules as 1 L of nitrogen or as 1 L of hydrogen.

The air of earth's atmosphere is primarily a mixture of nitrogen and oxygen. Small amounts of argon, carbon dioxide, and, especially on a humid day, water vapor, are also present. However, this text treats dry air as consisting simply of nitrogen and oxygen. Out of every 100 molecules of dry air, 79 molecules are nitrogen and 21 molecules are oxygen.

If dry air were separated into nitrogen and oxygen (by liquefying and distilling the air), 100 L of dry air at a given temperature and pressure would yield 79 L of nitrogen gas and 21 L of oxygen gas, at the same temperature and pressure. Therefore, dry air consists of 21 percent oxygen by volume.

Because the oxygen molecule is 32/28 times as heavy as the nitrogen molecule, dry air consists of more than 21 percent oxygen by weight. The following calculation shows that dry air consists of 23.3 percent oxygen by weight:

$$\frac{21 \cdot 32}{(21 \cdot 32) + (79 \cdot 28)} = 23.3$$

As noted, the molecular speed of any gas increases with increasing temperature. The average speed is proportional to the square root of the absolute temperature (K). Accordingly, the higher the temperature, the more violent are the collisions between gas molecules, and the greater is the likelihood that collisions will cause one or both of the collision partners to decompose. At very high temperatures (greater than about 5000 K or 6000 K), collisions are so violent that molecules can no longer exist, and a gas consists of only free atoms (some of which might be ionized) and free electrons.

Properties of Liquids

If the temperature of a gas is reduced, which reduces the molecular speed, or if the pressure is increased, which forces the molecules closer together, a point can be reached at which the gas condenses into a liquid. This is the point at which the attractive forces between the molecules overcome their tendency to separate after a collision.

In a *liquid*, the molecules are in contact with one another but also move relative to one another. For example, if a drop of ink is put into a glass of water, the ink will diffuse slowly through the water because of the molecular motion.

Every liquid has a *vapor pressure*, which increases with increasing temperature. The vapor pressure of a liquid is the pressure of the vapor over the liquid at which the rate of evaporation is equal to the rate of condensation; therefore, no net transfer occurs across the interface.

Vapor pressure can be expressed in millimeters of mercury (mm Hg) because a mercury column often is used to measure vapor pressure (760 mm Hg equals 1 atm, which equals 101.3 kPa). Consider liquid water, which has a vapor pressure of 17.5 mm Hg at 20°C. If water vapor at a pressure greater than 17.5 mm Hg exists above the liquid water, then the water vapor will condense. However, if water vapor is present at a pressure less than 17.5 mm Hg, or if no water vapor is present, then liquid water at 20°C will evaporate until the pressure of the water vapor reaches 17.5 mm Hg, or until the temperature of the liquid water drops below 20°C because of evaporative cooling.

If dry air exists over liquid water at 20°C, evaporation into the dry air will occur. If the dry air is at a pressure above 17.5 mm Hg, evaporation will occur slowly. However, if the pressure of the dry air is below 17.5 mm Hg, the water will boil, with rapid evaporation. If the water temperature is 100°C, boiling will occur unless the pressure of the dry air is greater than 1 atm.

Table 3.1 shows the vapor pressures of water and two combustible liquids for a series of temperatures. Additional data on vapor pressure can be found in various reference books, such as the *Handbook of Chemistry and Physics* [1] and *Perry's Chemical Engineers' Handbook* [2].

TABLE 3.1   Vapor pressures of three liquids at different temperatures

Temperature (°C)	Vapor pressure (mm Hg)		
	Water (H_2O)	Methanol (CH_3OH)	Hexane (C_6H_{14})
0	4.58	29.7	46.2
10	9.21	53.8	75.7
20	17.5	93.8	120
30	31.8	158	184
40	55.3	256	275
50	92.5	404	401
60	149	620	571
70	234	930	796
80	355	—	1091
90	526	—	1469
100	760	—	—
110	1075	—	—

Properties of Solids

When the temperature of a liquid is reduced, generally a freezing point will be reached at which the liquid changes into a crystalline solid. In a crystal, the atoms, ions, or molecules are fixed in regular geometric positions and cannot move through the solid. However, they can vibrate — move back and forth on either side of their equilibrium positions in the crystalline lattice. When crystals such as ice or sodium chloride are heated sufficiently, they melt. The melting point is the same temperature as the freezing point.

Some liquids do not have a sharp freezing point and do not form crystals upon cooling. Instead, these liquids become progressively more viscous as they are cooled, so that the molecules are less and less free to move about, and finally the substance entirely loses its capability of flowing and it becomes a solid. The liquids capable of such transitions generally consist of relatively large molecules. When they solidify, the molecules are trapped in a random arrangement, as contrasted with the orderly arrangement found in a crystal. Such solids are called amorphous substances or glasses; ordinary window glass is a common example. When window glass is heated, it gradually softens over a range of hundreds of degrees, rather than having a sharp melting point, as do crystals of ice and sodium chloride. Other examples of glassy solids include tar, asphalt, cold molasses, waxes, and many synthetic polymers such as polyvinyl chloride (PVC) and polymethyl methacrylate (Plexiglas™).

Further, the molecules might not be oriented completely randomly in some glassy substances, but might be aligned with each other to some degree, depending on the process by which the melt was solidified. In such cases, the degree of crystallinity of the solid can be determined by X-ray diffraction tests.

Metals are a special kind of solid. Metals generally consist of positively charged atomic ions in a geometrically defined crystal lattice, with electrons free to move through the lattice. The fact that metals conduct heat and electricity far better than other solids is due to the high mobility of the electrons through the crystal lattice.

A metal can consist of a single pure element such as copper, aluminum, iron, or 24-carat gold, or it can be an alloy of two or more elements. For example, brass is an alloy of copper and zinc, steel is an alloy of iron and carbon, and 18-carat gold is an alloy of gold and silver.

PHYSICAL CHANGE

For a pure substance, the transition from a gas to a liquid (or the reverse) depends on the prevailing temperature and pressure. High temperature, low pressure, or both favor the evaporation of a liquid, whereas low temperature, high pressure, or both favor the condensation of a gas or vapor.

Similarly, the transition between the liquid and the solid state (freezing or melting) of a pure substance depends on temperature and pressure. Low temperature always favors freezing, but high pressure can favor either freezing or melting, depending on the relative densities of the liquid and solid forms of the substance.

As is well known, ice floats on water, and hence must be less dense. High pressure favors the conversion of the less dense form, ice, to the more dense form, water. (This phenomenon is a consequence of *Le Châtelier's principle*.) Therefore, ice skating is possible because melting occurs under the skate blade, which provides lubrication as long as the temperature is not too far below 0°C. In contrast to water, crystalline benzene is denser than liquid benzene, and high pressure favors freezing rather than melting of benzene.

With a mixture of substances, the transitions between gas, liquid, and solid states will depend on the relative proportions of the substances as well as temperature and pressure. The principles governing these transitions are complex. For example, the freezing point of a mixture of ethylene glycol and water is much lower than that of either pure ethylene glycol or pure water. Similarly, vapor pressures of liquid mixtures follow complex laws and will not be discussed further in this book.

When a solid melts, an energy change is involved and heat must be added to the solid to convert it to a liquid. For example, 334 J of heat energy must be added to 1 g of ice at 0°C to convert it to liquid water at 0°C. This energy is called the *heat of fusion*.

Heat also is required to raise the temperature of a substance from a lower to a higher value when no change of state (phase change) is involved. For example, 4.18 J of heat energy are required to raise the temperature of 1 g of liquid water by 1°C. Therefore, the heat capacity of liquid water is 4.18 J/g per °C. The heat capacity of ice is 4.23 J/g per °C, while the heat capacity of water vapor is 2.03 J/g per °C at 100°C.

The following equation can be used to calculate the heat required to change the temperature of a given mass:

$$Q = m \cdot C \cdot \Delta T$$

where:

Q = heat required (J)

m = mass (g)

C = heat capacity of the substance (J/g per °C)

ΔT = temperature change (°C)

This expression is valid only if no change of state (e.g., from solid to liquid or from liquid to gas) occurs within the temperature range ΔT. Table 3.2 shows the heat capacity of some common substances.

To convert liquid water at 100°C to water vapor at 100°C, a heat of vaporization must be supplied. *Heat of vaporization* is the energy absorbed when a unit of mass of a liquid vaporizes. For water, this is 2257 J/g at 100°C. The heat of vaporization of hexane (mineral spirits) is only 342 J/g, and this explains why hexane evaporates much more readily than water. Handbooks such as *Handbook of Chemistry and Physics* [1] and *Perry's Chemical Engineers' Handbook* [2] provide values for heat capacities, heats of fusion, and heats of vaporization for hundreds of substances.

TABLE 3.2 Heat capacity of some common substances

State	Substance	Heat capacity* (J/g-K)
Gases	Air	1.1
	Water vapor	2.0
Liquids	Hexane	1.7
	Methanol	2.5
	Water	4.2
Solids	Copper	0.38
	Steel	0.46
	Gypsum plaster	0.84
	Aluminum	0.9
	Concrete	0.9
	Polystyrene	1.2
	Polyethylene	1.9
	Wood (oak)	2.4
	Wood (pine)	2.8

* These values are all approximate because heat capacity varies somewhat with temperature. The values for the gases are valid for heating at a constant pressure of 1 atm. The values for the liquid and solids are not appreciably dependent on pressure.

The *heat of condensation* of a vapor to a liquid has the same numerical value as the heat of vaporization of the liquid, but vaporization absorbs heat while condensation releases heat. Similarly, the heat of fusion (melting) of a solid is equal and opposite to the *heat of solidification* of the same substance in the liquid state.

It is also possible for a solid to vaporize directly, without any liquid forming. For example, a piece of ice in a cold vacuum chamber will gradually vaporize without melting. As another example, paradichlorobenzene crystals ($C_6H_4Cl_2$, used as a moth repellent) will, over a period of time, vaporize into air at room temperature. This process is called *sublimation*; the *heat of sublimation* of a solid is approximately equal to the sum of its heat of fusion and its heat of vaporization.

The converse of sublimation, the direct change from a vapor to a solid, can also occur. For example, water vapor in the atmosphere can change into snow or frost. When gaseous carbon dioxide (at high pressure) is discharged from a handheld extinguisher, the expansion causes cooling, and white particles of solid carbon dioxide form.

In summary, physical changes from state to state can occur between the three states of matter: gas, liquid, and solid. These are sometimes called phase changes. They always involve substantial amounts of energy absorbed or released. Going from a gas to a liquid to a solid releases energy, whereas going from a solid to a liquid to a gas absorbs energy.

CHEMICAL CHANGE

In a physical change, for example from a liquid to a vapor, there is no change in the molecules of which a substance is composed. The molecules are simply arranged differently and have different degrees of mobility in a gas, a liquid, a solution, a crystal, or a glass. However, it is possible for molecules to change into other molecules. The atoms themselves do not change, of course, but combine in different ways with other atoms, often forming new chemical bonds. Following are some examples of the many types of chemical change.

1. Dissociation: $H_2 \rightarrow 2\,H$

2. Association: $2\,H \rightarrow H_2$

3. Bimolecular exchange reactions
   a. $H_2 + Cl_2 \rightarrow 2\,HCl$
   b. $CO + OH \rightarrow CO_2 + H$
   c. $HCl + NaOH \rightarrow NaCl + H_2O$

4. Polymerization

   $2\,C_2H_4$ (ethylene) $\rightarrow\ C_4H_8$ (butylene)
   $C_4H_8 + C_2H_4\ \rightarrow\ C_6H_{12}$ (hexene)
   $C_6H_{12} + C_2H_4\ \rightarrow\ C_8H_{16}$ (octene)
   $C_8H_{16}\ \ldots\ \rightarrow$ polyethylene C_nH_{2n}, where n is large

5. Oxidation: In each of these oxidation reactions, a series of individual reaction steps is involved. However, only the overall reaction is shown.

   a. Hydrogen: $2\,H_2 + O_2 \rightarrow 2\,H_2O$
   b. Carbon monoxide: $2\,CO + O_2 \rightarrow 2\,CO_2$
   c. Methane: $CH_4 + 2\,O_2 \rightarrow CO_2 + 2\,H_2O$
   d. Propane: $C_3H_8 + 5\,O_2 \rightarrow 3\,CO_2 + 4\,H_2O$

6. Decomposition

    a. CH_4 (methane) $\rightarrow$ C (solid carbon) $+ 2 H_2$

    b. $(C_2H_4)_n$ (polyethylene) $\rightarrow n\ C_2H_4$ (ethylene)

The preceding list of chemical changes constitutes only a very small sample of the millions of chemical changes that can take place. However, all chemical changes have the following common features:

1. The reactions are reversible. That is, if, under certain conditions, the reaction

$$\text{reactants} \rightarrow \text{products}$$

can take place, then under other conditions the reverse reaction can take place:

$$\text{products} \rightarrow \text{reactants}$$

2. The molecules that make up the products are different from the molecules that make up the reactants. The products and reactants might have different colors, different odors, different melting and boiling points, different toxicities, and so on.

3. The rate at which any chemical change takes place depends on the temperature. The higher the temperature, the more rapidly the chemical change occurs. Many chemical systems are unreactive at room temperature but react very rapidly at higher temperatures. For example, a piece of acid-free paper exposed to air at room temperature for 100 years will undergo only a very slight degree of chemical reaction, which is evidenced by yellowing and brittleness. However, if this piece of paper is placed in a furnace containing air at 600°C, it will burst into flames and be consumed in a few seconds.

4. When a chemical change occurs, it is always accompanied by a change of energy. That is, either heat is released or heat is absorbed. The quantity of heat is always the same for a given chemical change (for example, $C + O_2 \rightarrow CO_2$) that occurs at a specified temperature and pressure.

Two additional general statements can be made about chemical changes when chemical reactions are involved in a fire.

1. Fires almost always involve oxidation reactions between various combustibles and the oxygen in the air. *(See Figure 3.1.)* These reactions release heat and are referred to as *exothermic reactions*.

FIGURE 3.1 Burning flammable liquid.

2. When liquids or solids are involved in a fire, the liquids vaporize first and the solids generally decompose or *pyrolyze* first, to produce vapors that then react with oxygen. The vaporization, decomposition, or pyrolysis processes almost always absorb heat. Reactions that absorb heat are called *endothermic reactions*. (Solid pyrotechnic mixtures and solid-propellant rocket fuels constitute exceptions to this rule; they decompose exothermically.)

PRINCIPLES OF COMBINING PROPORTIONS

Because fire consists of chemical reactions between gaseous, liquid, or solid combustibles and the oxygen in the air, it is useful to know the proportions in which these ingredients combine. As a simple example, consider the burning of hydrogen gas (which exists as H_2 molecules) by reacting with oxygen (which exists as O_2 molecules):

$$2 \ H_2 + O_2 \rightarrow 2 \ H_2O$$

The atomic weight of hydrogen is 1, while that of oxygen is 16. Thus, the molecular weight of hydrogen is 2, that of oxygen is 32, and that of water is 18 $(1 + 1 + 16)$. The combining proportions of the preceding reaction are

$$2(2) + (32) \rightarrow 2(18), \quad \text{or}$$

$$4 + 32 \rightarrow 36$$

Therefore, 4 g of hydrogen will combine with 32 g of oxygen to form 36 g of water.

Dry air is a mixture of 23.3 percent oxygen by weight and 76.8 percent nitrogen by weight; therefore 4 g of hydrogen will require $32/0.233 = 137$ g of dry air for complete combustion. Or, for each gram of hydrogen, $137/4 = 34.3$ g of dry air are required. If less air than this is available, only partial combustion can occur.

Often, however, it is necessary to know the volumes of gases rather than the weights of gases. Therefore, the concept of the mole in chemestry must be introduced. A mole of a chemical species is a quantity of that species equal to its molecular weight in grams. For example, a mole of hydrogen is 2 g of hydrogen $(2 \cdot 1)$. One mole of oxygen is 32 g of oxygen $(2 \cdot 16)$. One mole of water is 18 g of water $(1 + 1 + 16)$.

A very important and simple principle exists that applies to volumes of gases. At a given temperature and pressure, 1 mole of any gas occupies the same volume as 1 mole of any other gas. Furthermore, if different gases are mixed at the same temperature and pressure, the volumes are additive. This principle is not valid for gases at extremely high pressures or at temperatures close to their liquefaction points. However, for fires at atmospheric pressure, the principle is accurate. To illustrate the principle, the equation

$$2\ H_2 + O_2 \rightarrow 2\ H_2O$$

shows that 2 volumes of hydrogen gas, or 2 moles, will combine with 1 volume of oxygen gas, or 1 mole, to form 2 moles of water. Because atmospheric air contains 21 percent oxygen by volume, 2 volumes of hydrogen will require $1/0.21 = 4.76$ volumes of dry air. Or, each volume of H_2 requires $4.76/2 = 2.38$ volumes of air.

What is the volume occupied by the water formed in the above reaction, relative to the volume of the original hydrogen? This volume clearly depends on whether the water is in the liquid or the vapor state. Assuming that the reactants, hydrogen and air, are originally at 25°C and 1 atm, and further, assuming that a substantial quantity of heat is released, so that the products, water and nitrogen, are at 2000°C, then the water will clearly be in the form of vapor. (If the products were cooled to 25°C, then 97 percent of the water would be in the form of liquid.) To calculate the volume occupied by water vapor at 2000°C, relative to the volume of the original hydrogen at 25°C, refer to the perfect gas law (addressed earlier in this chapter). The perfect gas law shows that the density of a gas is inversely proportional to its absolute temperature (K). The volume of a given mass of gas is the inverse of its density. Accordingly, *at constant pressure the volume occupied by a given mass of gas is directly proportional to its absolute temperature.*

To apply this law to the question under discussion, convert 2000°C to 2273 K, and 25°C to 298 K. Then, assume that each volume of hydrogen at

298 K would be transformed to 1 volume of water vapor at 298 K (which, in fact, would not occur because of condensation), or to 2273/298 volumes of water vapor at 2273 K. That is, 2273/298 = 7.63 volumes of water vapor at 2000°C would form from combustion of 1 volume of hydrogen with 2.38 volumes of air.

Note that the 2.38 volumes of air would contain 0.79 · 2.38 = 1.88 volumes of nitrogen at 25°C, which would expand to 1.88 · 2273/298 = 14.34 volumes of nitrogen at 2000°C. On adding the 7.63 volumes of water vapor to the 14.34 volumes of nitrogen, the overall result is that 21.97 volumes of combustion products at 2000°C would form from 1 volume of hydrogen and 2.38 volumes of air at 25°C.

Now, consider a different combustion reaction, that of propane, C_3H_8:

$$C_3H_8 + 5\ O_2 \rightarrow 3\ CO_2 + 4\ H_2O$$

This equation shows 1 mole of propane combining with 5 moles of oxygen to form 3 moles of carbon dioxide (CO_2) and 4 moles of water. The numbers 5, 3, and 4 are required to balance the equation. There are 10 oxygen atoms on the left side of the equation (5 · 2) and 10 oxygen atoms on the right side of the equation (3 · 2 + 4 · 1). Also, there are 3 atoms of carbon on each side and 8 atoms of hydrogen on each side. Thus, the equation is balanced. The procedure for balancing such an equation is as follows:

1. Write $C_3H_8 + x\ O_2 \rightarrow y\ CO_2 + z\ H_2O$ (x, y, and z are unknown).
2. Assigning $y = 3$ makes the carbon balance correct.
3. Assigning $z = 4$ makes the hydrogen balance correct.
4. Then, knowing y and z, study the oxygen balance. The right side has $3 \cdot 2 + 4 \cdot 1$, or 10 oxygen atoms. Therefore, the left side must have 5 oxygen molecules O_2 (10 oxygen atoms), so x must be 5.

One mole of propane weighs 44 g (3 · 12 + 8 · 1). Five moles of oxygen (O_2) weigh 160 g (5 · 32). Therefore, each gram of propane requires 160/44 = 3.64 g of oxygen for complete combustion (or 3.64/0.233 = 15.6 g of air).

Now, consider the case of a solid, such as carbon, burning in oxygen:

$$C_{(s)} + O_2 \rightarrow CO_2$$

Here, 1 mole (12 g) of carbon (C) reacts with 1 mole (32 g) of oxygen (O_2) to form 1 mole (44 g) of carbon dioxide (CO_2). The subscript (s) denotes the solid state.

Now, suppose that insufficient oxygen is present to oxidize the carbon completely, and, as a result, some carbon monoxide (CO) as well as some carbon dioxide (CO_2) forms. Instead of 1 mole of oxygen per mole of carbon, assume that only 0.8 mole of oxygen ($0.8 \cdot 32 = 25.6$ grams) is available per mole (12 grams) of carbon. Then,

$$C_{(s)} + 0.8\ O_2 \rightarrow x\ CO + y\ CO_2$$

From the carbon balance,

$$1 = x + y$$

From the oxygen balance,

$$(0.8)(2) = 1.6 = x + 2y$$

Solving this pair of equations for x and y yields $x = 0.4$ and $y = 0.6$. Then, the reaction equation is

$$C_{(s)} + 0.8\ O_2 \rightarrow 0.4\ CO + 0.6\ CO_2$$

Now, assume that the oxygen came from atmospheric air, with 79 percent nitrogen by volume and 21 percent oxygen by volume. Recall the relationship between moles of gas and volumes of gas. Air consists of 79 mole percent nitrogen and 21 mole percent oxygen. Accordingly, in order to take the nitrogen into account, write the previous equation as

$$C_{(s)} + 0.8\left[O_2 + \left(\frac{79}{21}\right)N_2\right] \rightarrow 0.4\ CO + 0.6\ CO_2 + 3.01\ N_2$$

where $3.01 = (0.8)(79/21)$. This form of the equation is used to calculate the volume percent of any species, for example, carbon monoxide, in the fire products:

$$\text{Volume \% CO} = \frac{0.4}{0.4 + 0.6 + 3.01} \bullet 100 = 9.98\%\ CO$$

Finally, consider the combustion of polyethylene, which is a polymer. Ethylene, a gas, has the formula C_2H_4. Polyethylene, a solid, consists of long linear chains of ethylene units. Each polyethylene molecule contains thousands of C_2H_4 units. Assume that there are n ethylene units per polymer molecule and that the formula is $(C_2H_4)_n$.

Before considering the combustion of this polymer, first write the equation for combustion of the monomer C_2H_4:

$$C_2H_4 + 3\ O_2 \rightarrow 2\ CO_2 + 2\ H_2O$$

In this equation, the coefficients were chosen so that the elements C, H and O are balanced. The equation shows that 28 g of ethylene $(12 + 12 + 1 + 1 + 1 + 1)$ require 96 g of oxygen $(3 \cdot 32)$ for complete combustion.

Now, write the equation for polyethylene:

$$(C_2H_4)_n + 3n \ O_2 \rightarrow 2n \ CO_2 + 2n \ H_2O$$

This equation shows that $28n$ g of polyethylene require $69n$ g of oxygen and can form $88n$ g of carbon dioxide $(2 \cdot 44)$ and $36n$ g of water $(2 \cdot 18)$.

Notice that, because n occurs in each term, its value does not need to be known in order to calculate combining proportions. Obviously, for any value of n, 28 g of polyethylene require 96 g of oxygen and form 88 g of carbon dioxide and 36 g of water.

Use of these principles permits calculation, either on a mass basis or, for gases, on a volume basis, of the combining proportions in which any chemical substance reacts with oxygen, as long as the following conditions are met:

1. The chemical formula of the combustible, in terms of its elements, is known.

2. The products that are expected to form are known. For example, if sulfur is burned, does it form SO_2 or SO_3?

3. A sufficient amount of oxygen is present.

For a combustible containing carbon, hydrogen, oxygen, nitrogen, chlorine, and sulfur, the primary products of complete combustion are carbon dioxide (CO_2), water (H_2O), nitrogen (N_2), hydrogen chloride (HCl), and sulfur dioxide (SO_2). Of course, combustion generally is incomplete in actual fires, but the foregoing method of calculation (called *stoichiometry*) is still very useful.

In cases where insufficient oxygen is available for complete reaction (called *stoichiometric reaction*) to final products, it generally is not possible to calculate the output of incompletely oxidized species, such as carbon monoxide. Imagine a mixture of hydrogen and carbon monoxide burning under conditions such that insufficient oxygen is available to react completely with both combustibles. The reaction could form $H_2O + CO$, or $H_2 + CO_2$, or some combination of both, while still satisfying all the stoichiometric principles discussed in this section. Furthermore, other products, for example soot, often form in unpredictable quantities when hydrocarbon fires occur. The case where carbon was oxidized to a calculable mixture of CO and CO_2 with a deficient amount of oxygen (discussed earlier in this section) was an exception, which could be calculated because hydrogen was absent.

ENERGETICS OF CHEMICAL CHANGE

Every chemical change is accompanied by a change of energy, because the chemical energy of the reaction products is different from that of the reactants. This change usually is manifested in the form of heat energy, but also can include electrical energy (for electrochemical processes, as in batteries) or mechanical energy, where expansion or contraction or kinetic energy (motion) is involved (as in explosions). Consider the following simple chemical reaction:

$$C_{(s)} + O_{2(g)} \rightarrow CO_{2(g)}$$

$$\Delta H_{298} = -393.5 \text{ kJ/mole}$$

This equation shows that if solid carbon [$C_{(s)}$] at 25°C (298 K) reacts completely with gaseous oxygen [$O_{2(g)}$] at 25°C, and the resulting hot carbon dioxide gas is cooled down to the original temperature (25°C), then exactly 393.5 kJ of heat energy must be removed in the cooling process for each mole (12 g) of carbon consumed. The minus sign before 393.5 means that the reaction is exothermic; that is, heat is released to the surroundings.

Whenever ΔH is positive, this signifies that the reaction is endothermic; that is, heat is absorbed from the surroundings. The evaporation of water is endothermic; the condensation of steam is exothermic.

Now consider another chemical reaction, the oxidation or combustion of carbon monoxide (CO):

$$CO_{(g)} + \frac{1}{2} O_{2(g)} \rightarrow CO_{2(g)}$$

and

$$\Delta H_{298} = -283.0 \text{ kJ/mole}$$

This equation shows what happens when 1 mole of carbon monoxide (12 + 16 = 28 g) is completely oxidized by $^1/_2$ mole of oxygen (32/2 = 16 g) to form 1 mole of carbon monoxide. If the process occurs at constant pressure, all species being gaseous, 283.0 kJ must be removed to cool the carbon dioxide down to the original temperature of 298 K.

Why is constant pressure mentioned? Any reaction occurring in the open takes place at 1 atm of pressure, which is, of course, constant (except for small barometric fluctuations). However, a reaction can occur in a closed container, with an increase or decrease of pressure. In the preceding example, 1.5 moles of reactant gases (1 mole of CO and $^1/_2$ mole of O_2) combine to form 1 mole of CO_2. After the CO_2 is cooled to the original temperature, its volume would be only two-thirds of the original volume if the process occurred at constant pressure. Thus, the surrounding atmosphere must compress the gas to the smaller volume by doing mechanical work on it and adding energy to it. (In the equation above, this energy is about 1.24 kJ, which is less than 1 percent of the 283.0 kJ shown in the equation.) If the same process occurred at constant volume instead of constant pressure, the heat energy liberated would be 1.24 fewer kJ.

Chemists generally use the symbol ΔH_{298} to refer to the energy of the reaction products at 298 K minus the energy of the reactants at 298 K, when the process occurs at constant pressure. ΔH is called the *enthalpy change*. For a combustion reaction, the negative of ΔH, which is then a positive number, is called the *heat of combustion at constant pressure*.

The symbol ΔE_{298} (or sometimes ΔU_{298}) refers to the energy of the reaction products at 298 K minus the energy of the reactants at 298 K, when the process occurs at constant volume, and often is called the *internal energy change*. For fires occurring at atmospheric pressure, the focus is on ΔH rather than ΔE (or ΔU).

To illustrate another point, consider the combustion of the gas ethylene C_2H_4 at constant pressure:

$$C_2H_{4(g)} + 3\ O_{2g} \rightarrow 2\ CO_{2(g)} + 2\ H_2O_{(g)}$$

$$\Delta H_{298} = -1323 \text{ kJ/mole},\quad \text{or}$$

and

$$C_2H_{4(g)} + 3\ O_{2(g)} \rightarrow 2\ CO_{2(g)} + 2\ H_2O_{(l)}$$

$$\Delta H_{298} = -1411 \text{ kJ/mole}$$

Note that the quantity of heat liberated by this reaction depends on whether the water in the products after cooling to 298 K (25°C) is assumed to be in the form of gas or liquid. At 25°C and 1 atm, water will be primarily a liquid rather than a gas. This generality suggests that the larger value, 1411 kJ per mole, which is called the higher or *gross heat of combustion*, should be used.

In fires and other combustion processes, however, the water vapor in the products does not condense immediately because cooling is slow. Therefore, the smaller value, 1323 kJ/mole of ethylene, which is the lower or *net heat of combustion*, should be used. The difference in the higher and lower values is the heat of condensation of water, or 44 kJ/mole at 298 K. In the ethylene case, 2 moles of water, or 88 kJ, are involved per mole of ethylene.

Both higher (gross) and lower (net) heats of combustion of many substances are tabulated in handbooks. [1, 2, 3] Table 3.3 lists the net heat of combustion for some common substances that might be involved in fire.

It has been determined [4] that if the heat of combustion of a substance is expressed as kilojoules per gram of the air required by the principles of combining proportions, then the heat of combustion is nearly the same for most combustible substances. The value is approximately 3 kJ per gram of air required.

TABLE 3.3 Net heat of combustion for various gases, liquids, and solids

State	Substance	Net heat of combustion*	
		KJ/mole	kJ/g
Gases	Carbon monoxide, CO	283	10.1
	Ethylene, C_2H_4	1323	47.2
	Hydrogen, H_2	242	121
	Methane, CH_4	803	50.2
	Propane, C_3H_8	2044	46.4
Liquids	Acetone, CH_3COCH_3	1660	28.6
	Benzene, C_6H_6	3138	40.2
	Hexane, C_6H_{14}	3858	44.8
	Methanol, CH_3OH	639	19.9
	Octane, C_8H_{18}	5058	44.3
Solids	Carbon, C	394	32.8
	Cellulose, $(C_6H_{10}O_5)_n$	—	16.1
	Polyethylene, $(C_2H_4)_n$	—	43.3
	Polystyrene, $(C_6H_5CHCH_2)_n$	—	39.9
	Wood	—	16–19

* Values are based on the water in products as vapor rather than liquid. To calculate gross values of the heat of combustion, add 44 kJ times the moles of water formed per mole of combustible to the kJ/mole column, or 2.4 kJ times the grams of water formed per gram of combustible to the kJ/gram column.

CHEMICAL EQUILIBRIUM AND CHEMICAL KINETICS

If half a dozen assorted types of molecules were placed in a box and the temperature were raised, chemical reactions would take place in most cases. Ultimately, however, the reactions would be expected to cease because the atoms in the box would have rearranged themselves into the most stable combinations possible at the prevailing temperature and pressure. This final state is called *chemical equilibrium*. The time required to reach this final state is controlled by *chemical kinetics*.

It is possible to calculate, by *chemical thermodynamic* procedures described elsewhere [5], the molecular composition that any mixture at a given temperature and pressure will attain when chemical equilibrium is reached. It is much more difficult, but possible in some cases, to calculate the time needed for the process to reach equilibrium. If the reaction rate cannot be calculated, it is often possible to measure the rate in an experiment. Results of many such measurements have been tabulated.

To illustrate these important concepts, consider a mixture of 2 moles of hydrogen and 1 mole of oxygen. Knowing that the reaction

$$2\ H_2 + O_2 \rightarrow 2\ H_2O$$

can occur, it is not surprising to discover that when this system reaches chemical equilibrium (at 25°C and 1 atm), 99.999+ percent of the molecules present will be H_2O molecules rather than H_2 and O_2 molecules.

However, when we perform the experiment and mix hydrogen and oxygen at 25°C, nothing observable happens. Even if the mixture is stored for a year, only a tiny fraction of the molecules will have reacted. Clearly, the reaction rate of this reaction is extremely slow at 25°C. Now, if the hydrogen–oxygen mixture is heated in a 7-cm diameter sphere, the mixture would explode suddenly at 560°C, changing completely to water within a fraction of a second.

If the mixture had been heated to only 500°C, at 1 atm, no reaction would have occurred. If the *pressure* then had been raised gradually, at 500°C, an explosion would have occurred at 4 atm. What is more surprising, if the pressure had been lowered at 500°C instead of being raised, an explosion would have occurred when the pressure had dropped to $1/20$ of an atmosphere.

This reaction has been studied extensively and is now understood thoroughly. In fact, a Nobel prize was awarded in 1956 to the scientists who first unraveled this mystery. So-called *chain reactions* of the free atoms and radicals H, O, OH, and HO_2 are involved. The rate at which any mixture of hydrogen and oxygen (or air) will go to equilibrium (that is, form water) can now be predicted accurately for any set of experimental conditions.

Similar chemical kinetic considerations apply to the ignition temperature of any combustible with air, except that the details of the chemical reactions often are quite complicated and in many cases have not yet been worked out fully.

A general finding in chemical kinetics is that rates of chemical reactions increase with increasing temperature. In some extreme cases, reaction rates double with every 10-degree increase in temperature. This change is called *exponential dependence*. The explanation for this behavior is that molecules move faster at higher temperatures and, as a consequence, undergo more violent collisions with one another, causing chemical bonds to break. This phenomenon is illustrated in Figure 3.2.

If a gas mixture is at a very high temperature, such as in a hot flame, then chemical reactions occur in a small fraction of a second; chemical equilibrium will be reached rapidly when the temperature is above about 2000 K.

Consider now a mixture of 2 moles of hydrogen and 1 mole of oxygen at 2500 K and 1 atm. Chemical equilibrium will be reached in a millisecond or less, and the primary product will be water. However, a temperature of 2500 K is so high that the H_2O molecules are not completely stable, and some of their decomposition products will be present at equilibrium. A thermodynamic calculation yields the following product distribution at 2500 K and 1 atm:

H_2O:	90.0%	O_2:	1.6%
H_2:	4.3%	H:	0.53%
OH:	2.3%	O:	0.18%

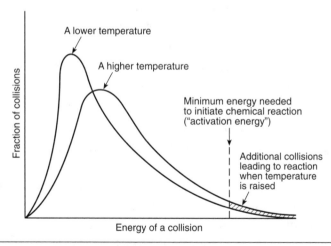

FIGURE 3.2 The relationship between the temperature of a gas and the fraction of molecular collisions capable of breaking chemical bonds and, therefore, causing a reaction to occur.

Most combustibles, when burned with just the stoichiometric amount of air under conditions where no heat is lost, will produce flames at temperatures from 2100 to 2300 K, and chemical equilibrium will be achieved in such flames.

In fires, however, combustion often occurs with a yellow luminous flame because of tiny, hot carbon particles that form and radiate heat energy. As much as 30 to 40 percent of the heat of combustion can be lost from such a flame because of this radiation. Consequently, the flame temperature often will be below 2000 K, and the combustion gases will become diluted with air and cooled before chemical equilibrium can be reached. This phenomenon can cause dangerous concentrations of carbon monoxide, even if an excess of oxygen is present around the flame.

Problems

1. What is the molecular difference between a gas, a liquid, and a solid?

2. If liquid water is in contact with humid air at 30°C, how high must the percentage by volume of water vapor in the air be to prevent evaporation?

3. How many joules of heat are needed to raise the temperature of 1 kg of gypsum plaster from 20°C to 150°C?

4. When water vapor condenses to a liquid, must heat be added or removed in order to maintain a constant temperature?

5. The properties of a material will change when it undergoes either physical change or chemical change. What is the difference between physical and chemical change?

6. What is the usual effect of temperature on the rate of chemical change?

7. If the temperature of 100 g of oxygen gas is raised from 20°C to 200°C, while the pressure is constant at 1 atm, how much does the gas expand relative to its initial volume?

8. How many grams of carbon dioxide can be produced from the combustion of 1 g of methane?

9. Explain the following terms: isotope, isomer, stoichiometric, mole, endothermic, net heat of combustion, and gross heat of combustion.

10. Balance the equation for benzene (C_6H_6) reacting with oxygen, to form carbon dioxide and water vapor.

11. What is the percentage of oxygen in dry air by volume? By weight? Why are the percentages different?

12. What are free atoms and free radicals? What is their relevance to fires?

References

1. *Handbook of Chemistry and Physics,* 78th ed., CRC Press, Boca Raton, FL, 1997.

2. Perry, R. W., and D. W. Green, eds., *Perry's Chemical Engineers' Handbook,* 7th ed., McGraw-Hill, New York, 1997.

3. Cote, A. E., ed., *Fire Protection Handbook,* 18th ed., National Fire Protection Association, Quincy, MA, 1997.

4. Huggett, C., "Estimation of the Rate of Heat Release by Means of Oxygen Consumption Measurements," *Fire and Materials,* 1980, Vol. 4, 1980, pp. 61–65.

5. Friedman, R., "Chemical Equilibrium," *The SFPE Handbook of Fire Protection Engineering,* 2nd ed., National Fire Protection Association, Quincy, MA, 1995, pp. 1-88–1-98.

CHAPTER 4

Flow of Fluids

LAWS GOVERNING MOTIONS OF A RIGID BODY

A review of the laws that govern the motions of a rigid body is presented before we consider the laws governing the flow of a fluid.

Acceleration of a Rigid Body

Consider an object of mass m, either at rest or moving at constant velocity v. The laws of motion, first stated by Newton, say that *a body will remain at rest or will continue moving at the velocity v unless it is acted upon by a net force.* The body is said to have a momentum equal to mv, the product of its mass and its velocity. Another way of expressing this law is to say that the body will *conserve its momentum*; that is, the action of an external net force is required to cause the momentum to change.

Now, if a net force F does act on a body of mass m, the body will accelerate at rate a in the direction of the applied force. (Acceleration is defined as the rate of change of velocity.) The acceleration rate a is directly proportional to the magnitude of the force F and is inversely proportional to the body's mass m. Written as an equation,

$$a = \frac{F}{m} \qquad (4.1)$$

A proportionality constant would be expected to be present in Eq. (4.1), but a judicious choice of units in which a, m, and F are expressed can result in the proportionality constant being unity and, hence, ignorable. One such set of units expresses acceleration in meters per second per second (m/s/s), force in newtons (N), and mass in kilograms (kg). (See Chapter 1 for definition of the newton and for conversion from newtons to pounds of force.) This set of units will be used in subsequent equations, and, for consistency, it will be necessary to express length in meters (m), velocity in meters per second (m/s), and time in seconds (s).

An alternative set of units resulting in a proportionality constant of unity in Eq. (4.1) is dynes, grams (g), centimeters (cm), and seconds (s). One newton equals 100,000 dynes.

The Effect of Gravitation on a Rigid Body

Any object of mass m, anywhere on the surface of the earth, is constantly acted upon by a downward gravitational force of about 9.8 N/kg of mass (or about 980 dyn/g of mass). The force is a few tenths of one percent weaker at the top of Mt. Everest than at sea level, but such small variations may be ignored for most engineering calculations.

Suppose the object is stationary, hanging from a cord. The gravitational force is exactly balanced by a tensile force in the cord, so there is no net force on the object, and it remains stationary.

Now suppose that the cord is cut at time t_0. The object will accelerate downward with an acceleration g, and its increasing velocity v may be calculated as

$$v = g(t - t_0) = \left(\frac{F}{m}\right)(t - t_0) \tag{4.2}$$

where g is 9.8 m/s/s, or 9.8 N/kg; velocity and time are expressed in meters per second (m/s) and seconds (s), respectively.

Thus the velocity increases linearly with time and may be calculated at any time interval. [Eq. (4.2) assumes that the frictional resistance force caused by the motion of the falling object through air is negligible. This is a good approximation for the first 1 or 2 seconds of fall. For an accurate description of the motion after this, a more elaborate equation may be developed by the use of calculus.]

The distance d the object has fallen during the interval $(t - t_0)$ may be calculated. Since the velocity increases linearly with time, the final velocity being v_f, the average velocity v_{av} is $v_f / 2$.

Therefore,

$$v_{av} = \frac{v_f}{2} = \frac{d}{t - t_0} \qquad (4.3)$$

Elimination of v_f between Eqs. (2) and (3) gives

$$d = 9.8 \frac{(t - t_0)^2}{2} \qquad (4.4)$$

where d is in meters and t is in seconds.

Potential Energy and Kinetic Energy: Mechanical Work

Potential energy is energy that is stored in an object. This energy may be mechanical, electrical, chemical, and so on. As one example, an object at a high elevation has potential energy relative to a similar object at a lower level. This difference in energy may be measured by the mechanical work that must be done to raise the object from the lower to the higher level. And, conversely, the object can be made to perform mechanical work while it descends from the higher to the lower level. The principle of *conservation of energy* specifies an equivalency between the change of potential energy and the mechanical work done on or by the object, as long as no other forms of energy (for example, friction that generates heat, kinetic energy) are involved.

Mechanical work, expressed in joules (J) or ergs, consists of a force F acting through a distance d; or it can be expressed as a pressure P, acting through a volume V. One J of work is done by a force of 1 N acting through a distance of 1 m, or a pressure of 1 N/m^2 acting through a volume of 1 m^3. [One erg of work is done by a force of 1 dyn acting through 1 cm. One J equals 10 million (10^7) ergs.]

The force acting on an object in a gravitational field is g times the mass of the object. Let h be the height (vertical distance) between two locations. Then, the difference in potential energy of an object of mass m at these two locations is given by

$$\Delta E_p = gmh \qquad (4.5)$$

Another form of energy is *kinetic energy*, associated with motion. Suppose that a force F acts on a body of mass m, initially at rest. The force acts while the object moves a distance d, accelerating all the while. (Assume that there is no other force acting, such as friction.) Then, the kinetic energy imparted to the object is Fd, which is the same as the mechanical work done on the object, according to the principle of conservation of energy.

Further, $F = ma$, by Eq. (1), and

$$d = \frac{at^2}{2} \qquad (4.6)$$

by a generalization of Eq. (4.4). Since the acceleration is constant,

$$v = at \qquad (4.7)$$

If time, t, is eliminated between Eqs. (4.6) and (4.7), the result is

$$d = \frac{v^2}{2a} \qquad (4.8)$$

Thus, from Eqs. (4.1) and (4.8), the increase in kinetic energy is given by

$$\Delta E_k = Fd = (ma)\left(\frac{v^2}{2a}\right) = \frac{mv^2}{2} \qquad (4.9)$$

Note that it is better not to think of either kinetic or potential energy as an absolute quantity. Rather, we should refer to changes of kinetic or potential energy relative to some convenient reference state.

Suppose that an object undergoes a change from kinetic energy to an equivalent quantity of potential energy, or vice versa. Since kinetic and potential energy are quantified by Eqs. (4.9) and (4.5), respectively, it follows that

$$\frac{mv^2}{2} = gmh \qquad (4.10)$$

Rearrangement of Eq. (4.10) to solve for v gives

$$v = \sqrt{2gh} \qquad (4.11)$$

Thus, if air resistance is neglected, an object dropped from height h will achieve a velocity v upon impact, independent of its mass. The velocity is proportional to the square root of the distance the object falls.

Conversely, an object projected upward with velocity v will reach a height h when it reaches its apogee (maximum height), calculable from Eq. (4.11).

BASIC ELEMENTS OF FLUID BEHAVIOR

The foregoing relations apply to a rigid solid. However, they may be modified to apply to a fluid, either gas or liquid.

For a fluid, the concept of pressure P (force per unit area) is used instead of force, and the concept of density ρ (mass per unit volume) is used instead of mass. The pressure may vary from one region in a fluid to another. For incompressible fluids such as water, the density is the same throughout. However for gases, which are readily compressible, the density can vary from one region of a fluid to another. (A relation between gas density and the pressure, temperature, and molecular weight is given by the *perfect gas law*, presented in Chapter 3.)

Consider now a vertical column of incompressible fluid of height h, cross-sectional area A, and density ρ. The gravitational pull on the column squeezes the fluid near the bottom of the column. As a result, the pressure within the fluid is higher at the bottom and decreases with increasing height. This pressure difference is easily calculated.

The mass of the column is equal to $hA\rho$. The gravitational force acting on this mass is equal to $ghA\rho$. The difference in pressure at the bottom of the column relative to the top of the column is ΔP. But, since pressure is force per unit area, the force difference associated with the pressure at top and bottom is $A\Delta P$. Now, since the column of fluid is not moving, and no other forces are acting, the pressure force must balance the gravitational force. When these two forces are equated, the cross section, A, cancels, and the result is

$$\Delta P = gh\rho \tag{4.12}$$

This important equation, while derived for an incompressible fluid, is also approximately true for air at atmospheric pressure, as long as no temperature gradients are present, and as long as the column of air is no more than a few meters high. The reason is that normal air has a very low density; gravitation causes only a 0.01 percent difference in pressure from top to bottom of an air column 10 m high.

Next, consider a fluid that is contained in a vessel, an orifice being present somewhere in the wall or bottom of the vessel. The pressure of the fluid inside the vessel, near the orifice, is higher than the external pressure by an amount ΔP. A jet of fluid will emerge from the orifice with velocity v. This velocity may be calculated as follows.

If friction is neglected, the potential energy of the stationary fluid inside the vessel is converted to kinetic energy, with no loss of energy. Consider a unit volume of fluid. Its mass is simply ρ. By Eq. (4.9), the kinetic energy of the unit volume of fluid after it accelerates to velocity v is $\rho v^2/2$. The potential energy that the unit volume of fluid had before emerging is equal to the work that would have to be done to push a unit volume of fluid back into the vessel against the adverse pressure gradient. This work is equal to the pressure difference times the volume of fluid (taken as unit volume), so the potential energy is ΔP. Equating potential and kinetic energy

$$\Delta P = \rho \left(\frac{v^2}{2}\right) \tag{4.13}$$

This equation may be rearranged in the form

$$v = \sqrt{\frac{2\Delta P}{\rho}} \tag{4.14}$$

Equation (4.14) shows the dependence of the fluid velocity on the square root of the excess of the fluid pressure over the external pressure. Either this excess fluid pressure could arise from the effect of gravitation on a column of fluid (hydrostatic effect), as given by Eq. (4.12), or it could arise by the action of a pump. Another possibility is that the excess pressure could be caused by thermal expansion of the fluid. There could be a combination of several of these effects.

First, assume that the pressure is solely hydrostatic. Then, Eq. (4.12) may be substituted into Eq. (4.14) to give

$$v = \sqrt{2gh} \tag{4.15}$$

Notice that density has canceled out. Also, notice that Eq. (4.15), for a fluid jet shooting from an orifice, is exactly the same as Eq. (4.11), for the velocity achieved by a falling rigid body in the absence of friction. The reason, of course, is that both equations are statements of the equivalence of kinetic energy as measured by velocity and potential energy associated with position in a gravitational field.

Now consider the case of fluid flow with a pump generating pressure; the fluid is also flowing from a higher to a lower level (or vice versa) and also has kinetic energy associated with its velocity. (However, friction is still being considered negligible, which is often a good approximation.) The fluid is incompressible. Assume that the fluid element is moving along a streamline, from point 1 to point 2. Then, *Bernoulli's equation* is valid:

$$P_1 + gh_1\rho + \rho(v_1)^2 = P_2 + gh_2\rho + \rho(v_2)^2 \tag{4.16}$$

The first term represents the energy from the pump, the second term the potential energy associated with position in the gravitational field, and the third term the kinetic energy. This Bernoulli equation says that the sum of these three energies is constant along a streamline, as long as negligible energy is dissipated by friction.

Note that Eq. (4.16) reduces to Eq. (4.13) when the flow is horizontal ($h_1 = h_2$) and reduces to Eq. (4.15) when there is no pump and the pressure is solely hydrostatic.

The meaning of flow "along a streamline" is clarified in the subsection Laminar and Turbulent Flow.

Viscosity

Imagine two coaxial cylinders, one within the other. A fluid occupies the annular gap between the two cylinders. The inner cylinder is mounted on frictionless bearings. A force, applied to this inner cylinder, causes it to rotate, while the outer cylinder is fixed in place. The force is needed to overcome the frictional resistance offered by the fluid.

Every fluid, gas or liquid, has a property called *viscosity*, which is a measure of the resistance it offers to a shearing force, as in the coaxial cylinder example. A notably viscous fluid will offer much more resistance than an airlike fluid.

Viscosity may be quantified as follows. Experiments show that the force F needed to maintain a velocity v of the inner cylinder is usually found to be proportional to this velocity v and also proportional to the surface area A of the inner cylinder. It is usually found to be inversely proportional to the width x of the gap between the two cylinders. Accordingly, an equation might be written

$$F = \mu\left(\frac{Av}{x}\right) \tag{4.17}$$

Here, μ, the proportionality constant, is the viscosity of the fluid in the gap. Its value may be obtained from an experiment in which F, A, v, and x are measured. Then, other experiments may be done, with different values of v, x, or both; if the viscosity turns out to be the same value, the fluid is said to be newtonian. Most common fluids, including all gases and water, are newtonian.

For a given fluid, the viscosity will vary with temperature but is only very slightly dependent on pressure. Table 4.1 shows values of viscosity, μ for a number of fluids.

Viscosity is important in fluid flow for three reasons:

TABLE 4.1 Viscosities of some fluids

Fluid	Temperature (°C)	Viscosity (N-s/m^2)
Air	20	0.0000183
Air	500	0.0000361
Air	1000	0.0000482
Water	0	0.001787
Water	20	0.001002
Water	90	0.0003147
Heptane	20	0.000409
Glycerin	20	1.49

1. When a fluid flows over an immersed object or through a duct, a drag force exists, the magnitude of which depends on the viscosity of the fluid. This arises when we consider the flow of water through a pipe or the propulsion of an airplane fuselage through air.

2. If we want to calculate the velocity at each point in a flow field, we must recognize that the flow will include a slow-moving boundary layer adjacent to any fluid-boundary wall or near any surface of an immersed object. The thickness of this boundary layer depends on viscosity. Outside the boundary layers, the flow is independent of viscosity and is governed by the principles of conservation of energy and momentum, discussed earlier in this chapter. The rate of heat transfer from the fluid to the wall depends on the thickness of the boundary layer.

3. As is discussed in the next section, a fluid may flow either in a laminar (streamlined) or turbulent (fluctuating) fashion, and viscosity is important in determining which mode occurs. Viscosity tends to suppress turbulence.

Each of these three effects of viscosity on fluid flow can be correlated by the dimensionless *Reynolds number*, $dv\rho/\mu$, where d is a characteristic distance and the other symbols are as previously defined.

The Reynolds number may be understood as being the ratio of the inertial force associated with the moving fluid to the viscous force restraining the motion, that is,

Inertial force ~ $\rho v^2 A$ [cf. Eq. (4.13)]
Viscous force ~ $\mu Av/d$ [cf. Eq. (4.17)]

Various formulas are presented later for pressure drop and boundary layer thickness as functions of the Reynolds number, under conditions of either laminar or turbulent flow. *(See Chapter 5, "Heat Transfer.")* The criterion for transition from laminar to turbulent flow, for a given geometry, can be expressed as a critical Reynolds number.

Laminar and Turbulent Flow

Consider the steady flow of a fluid at average velocity v being pumped through a pipe of internal diameter d. Regardless of whether the fluid is air, water, or glycerin, the flow will be steady, or *laminar,* if the Reynolds number is less than about 2000. The flow will be *turbulent*, or fluctuating, if the Reynolds number is more than about 2500. If the velocity is gradually increased through this range, the exact point of transition will depend on the following properties:

1. The smoothness or roughness of the inner wall of the pipe

2. The presence or absence of any pressure fluctuations in the flow, perhaps caused by the pump

Figure 4.1 shows some differences between laminar and turbulent flow in a pipe. The diagrams show an ink tube injecting a tracer downstream at the centerline. Under laminar flow conditions, the tracer follows the central streamline, with very little lateral spread. In contrast, the lateral spread is very rapid under turbulent conditions, and the fluctuating nature of the flow is visually evident.

Note that the velocity profiles are quite different for the two kinds of flow. In the laminar case, the average velocity is exactly half of the peak velocity. For the turbulent case, the average velocity is about 80 percent of the peak velocity. The viscous drag at the wall is much greater under turbulent conditions. The pressure drop required to force the flow through the pipe, at a Reynolds number of 2300, is almost twice as great under turbulent compared to laminar conditions.

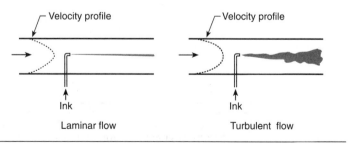

FIGURE 4.1   Laminar pipe flow and turbulent pipe flow.

Next, consider another geometry: a uniform laminar flow passing over a sharp-edged flat plate, as shown in Figure 4.2. A laminar boundary layer begins to form at the leading edge. (The boundary layer may be defined as the region where the velocity is less than 99 percent of the free stream velocity v.) The thickness of the boundary layer increases with the square root of the downstream distance from the leading edge, as long as the boundary layer remains laminar. At a given position, the thickness of the boundary layer in the laminar region is proportional to the square root of the viscosity and inversely proportional to the square root of the free stream velocity.

Let x, the downstream distance, be the characteristic distance in the Reynolds number $xv\rho/\mu$. At a downstream position such that the Reynolds number defined this way exceeds about 500,000, a transition to turbulence occurs. The frictional drag on the flat plate is much greater in the region where turbulence is present.

Consider yet another case. Hot air is generated by a local heat source (perhaps a fire), and rises, because of buoyancy. If the heated region is small, like a candle flame, the upward flow will be laminar. However, if the heated region is as wide as 0.5 m, the upward flow, referred to as a plume, will be turbulent. A turbulent plume will entrain surrounding air much more rapidly than a laminar plume; this has a major influence on the centerline temperature, velocity, and smoke concentration (in the case of a fire).

The value of a dimensionless number, called the *Grashof number*, is the guide to whether a buoyant plume will be laminar or turbulent. The Grashof number represents the ratio between the buoyant force and the viscous force acting on the fluid. It is proportional to the cube of the characteristic dimension (the width) of the plume.

Another common way for turbulence to be generated is for a laminar flow to encounter an object, so that the fluid must flow around the object. If the Reynolds number is high enough (depending on the shape of the object, especially in the absence of streamlining), a turbulent wake will form downstream of the object.

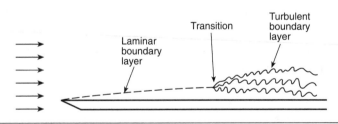

FIGURE 4.2 Transition from laminar to turbulent flow in a boundary layer on a sharp-edged flat plate.

Characteristics of Turbulence

While turbulence is very important in the study of fluid flow, it is not fully understood. It is difficult to predict from first principles when a transition will occur from laminar to turbulent flow, for some selected geometry. Given that a flow field is turbulent, there are various degrees of turbulence that may be present. The frictional characteristics of the flow (the effective viscosity) and the rate at which heat is transferred through the fluid depend greatly on the nature of the turbulence.

Turbulence may be better understood by thinking of it as possessing two characteristics:

1. Intensity
2. Scale

Since turbulence consists of more or less random fluctuations, the intensity will be some sort of average of intensity measured at various nearby locations at a given time, or an average measured over time at a given location. Similarly, the scale will be an average of scales over a period of time or a range of locations.

In certain cases, the turbulence intensity and scale may not change as one moves locally through the fluid in any direction. This type of turbulence is called isotropic turbulence. However, for turbulent flow near a wall, for example, the variation of turbulent properties in the direction normal to the wall will be quite different from the variation in the flow direction.

Turbulent intensity may be defined as some sort of average of the fluctuating component of the flow velocity in any chosen direction. It is usual to define it as the square root of the average of the square of the fluctuating component of the velocity in a given direction. This apparently strange definition is chosen because the kinetic energy of a fluctuating "clump" of fluid is proportional to the square of its velocity v. *[See Eq. (4.9).]* This relates the turbulent intensity to the kinetic energy associated with turbulence.

Turbulent scale is based on the correlation between the turbulent intensity at two nearby points in a turbulent flow, at a given instant. If the two points are very close together, the correlation will be nearly perfect (correlation of unity); and if the two points are substantially far apart, there will be essentially no correlation (zero correlation). The scale of the turbulence might be defined as the separation distance for which the correlation is 0.5.

Instruments have been developed to measure the intensity and scale characteristics of turbulence. One such instrument is a hot-wire anemometer. Another is a laser-Doppler velocimeter (LDV).

One way to produce isotropic turbulence is to cause a laminar flow to pass through a coarse grid or a perforated plate with sufficient velocity so that the Reynolds number based on the thickness of a grid element is greater than a critical value. Then, isotropic turbulence will exist downstream of the grid, with a scale related to the grid dimensions. The intensity of the turbulence may be as little as one percent of the average velocity, or in excess of 25 percent, depending on the degree of flow blockage by the grid. As this turbulent flow continues downstream from the grid, the turbulence will gradually decay; that is, its intensity will diminish.

Now consider the turbulent flow in a pipe, which is nonisotropic. The scale of the turbulence at the centerline is relatively large and is much smaller near the walls. However, the turbulent intensity is greater halfway from the center to the wall than at the center.

The rate of turbulent diffusion of either a trace species or of heat through the fluid is proportional to either the concentration gradient or the temperature gradient. The proportionality constant in this relationship is equal to the product of the intensity and the scale of the turbulence.

Calculation of Flow Patterns

A major task in fluid mechanics is the calculation of flow patterns. It is assumed that the fluid is confined by rigid surfaces of known geometry and is acted upon by a pressure, which might be caused by a pump, by gravitational force, or by thermal expansion. The density and viscosity of the fluid are known.

Such a situation is describable by a set of equations that are based on the principles that mass, energy, and momentum are conserved and that the fluid obeys an equation of state, such as the perfect gas law. If the flow is laminar, there are some cases where it is possible to solve the equations. In other cases, even under laminar conditions, analytical solution of the equations is not possible, but the equations may be solved numerically with a high-speed computer.

The computer solutions involve dividing the space of interest into numerous small regions called *cells*. As many as several hundred thousand cells may be used in a computation. The computer calculates how the fluid moves from each cell to the adjacent cells, at each instant. Tens of millions of calculations are needed.

For laminar flows, this procedure will generally give correct solutions, as long as the grid is fine enough. When the flow is turbulent, which is often the case, the same numerical approach is used, but the problem becomes much more difficult. The computer power needed to calculate every fluctuating motion of every particle of the fluid is far beyond what is now available, and furthermore, one is generally not interested in this kind of detail but rather in the overall behavior of the fluid. Several approximate methods have been developed to compute turbulent flow, but they all involve approximations of one sort or another. The results of these calculations are not completely accurate, but they often give results reasonably close to the actual fluid behavior, as observed.

In many cases, analytical or numerical calculations of flow are not necessary, because these cases either have already been calculated or have been studied experimentally, for instance in wind tunnels, and the results are available in formulas or graphs, often involving dimensionless variables.

The foregoing presentation of fluid flow is only an introduction; the reader who wishes to learn more is referred to *The SFPE Handbook of Fire Protection Engineering* [1] or to a textbook on fluid mechanics, such as that by Kleinstreuer [2].

Problems

1. What is the velocity of a freely falling object weighing 3 kg, 2 s after its release?

2. What is the kinetic energy of the object in Problem 1?

3. How far will the object in Problem 1 have fallen?

4. An open tank contains water (density 1000 kg/m^3) 3 m deep. What is the pressure in the water at the bottom of the tank, relative to atmospheric pressure?

5. A hole is drilled in the bottom of the above tank. What is the velocity at which a water jet issues from the hole?

6. Why is the presence or absence of turbulence in a flowing fluid important?

References

1. *The SFPE Handbook of Fire Protection Engineering*, 2nd ed., National Fire Protection Association, Quincy, MA, 1995.

2. Kleinstreuer, C., *Engineering Fluid Dynamics*, Cambridge University Press, Cambridge, England, 1997.

CHAPTER 5

Heat Transfer

Heat transfer is an energy flow from one location to another because of a temperature difference. Heat always flows from a higher to a lower temperature region. Before we discuss how this occurs, we must understand the nature of temperature and of heat.

Temperature is an intensive property of an object, and heat is an extensive property. This means that the temperature of an object is not related to the mass of the object, whereas the heat (the thermal energy) contained in an object at a given temperature is directly proportional to the mass of the object. Pressure, density, and concentration are also intensive properties, and mass and volume, in addition to energy, are extensive properties.

Heat, or thermal energy, is a form of energy, as are kinetic energy, potential energy associated with position in a gravitational field, electrical energy, chemical energy, and so on. Energy is conserved when one form is converted into another. Heat is contained in matter, that is, in assemblies of atoms or molecules. In a monatomic gas, the heat content is simply the sum of the kinetic energies of the individual atoms. The kinetic energy of an atom or molecule is proportional to the product of its mass and the square of its velocity. [See Eq. (4.9).] The temperature of a group of monatomic gas atoms is proportional to the kinetic energy of an average atom in the group. If there are twice as many atoms with the same average velocity, then the heat content is twice as much but the temperature is unchanged.

A polyatomic gas has not only kinetic energy, but also energy stored within each molecule, as vibrational and rotational energy. Hence, at the same temperature, a given number of molecules of a polyatomic gas will contain significantly more heat than the same number of molecules of a monatomic gas. (The heat content of any substance is zero at the zero point of absolute temperature.)

Since the molecules in a solid are fixed in place, they cannot possess kinetic energy. However they can vibrate, and they do so. The higher the temperature, the greater the amplitude of the vibrations.

In summary, the temperature of a substance is proportional to the kinetic, vibrational, and rotational energies of an average atom or molecule of the substance; whereas the heat content, or thermal energy, of a substance is equal to the sum of the kinetic, vibrational, and rotational energies of all its constituent atoms or molecules.

MODES OF HEAT TRANSFER

Heat moves through a medium in two ways: through *conduction* and through *radiation*. (Convection, discussed later in this section, involves conduction through a moving fluid.) The information on conductive, radiative, and convective heat transfer, as presented in this text, is intended to be introductory; refer to specialized texts such as *The SFPE Handbook of Fire Protection Engineering* [1] for a more detailed review.

1. *Conduction.* Imagine two objects having different temperatures and suddenly being placed in contact with each other. The atoms or molecules in the higher-temperature object will have higher energies, on the average. Some of this energy will flow from the hotter to the colder object. If the two objects are solids, the larger amplitude vibrations of the atoms or molecules in the hotter object will pass over some of their energy to the neighboring colder atoms or molecules. If, instead, a hot and a cold gas are brought into contact, then the energy transfer is a little different. The mobile hotter and colder molecules will interpenetrate one another (diffusion). Then, when a hot (higher energy) molecule finds itself surrounded by lower energy molecules, it will transmit some of its energy to them by undergoing collisions.

2. *Radiation.* The other basic mode of heat transfer is radiation. This mode does not require molecules to be in contact with one another. Indeed, radiation can occur through a vacuum. Note that the sun's radiation reaches the earth through the vacuum of space. Radiation can also pass through any transparent medium such as dry air or a semi-transparent medium such as water or glass.

Radiation has a spectrum of wavelengths or frequencies, ranging from X rays to radio waves. The part of the radiation spectrum of greatest interest in the transfer of heat is the infrared region.

The basic principle, in summary, is that every object at any temperature above absolute zero is always emitting radiant energy to its surroundings, and also absorbing radiant energy from its surroundings. The higher the temperature, the greater the rate at which this energy is emitted. In fact, the radiant emission is proportional to the fourth power of the absolute temperature of the emitter. Thus, radiative transfer is very important when high temperatures are involved.

Fourier's Law of Heat Conduction

Heat flows through a material in the direction from the high-temperature region to the low-temperature region. The rate of heat flow is proportional to the temperature gradient. This is expressed mathematically by the *Fourier equation* for steady-state heat flow:

$$q' = kA\left(\frac{\Delta T}{\Delta x}\right)$$ (5.1)

where:

q' = the rate of heat flow (J/s or watts) in the x direction

A = the area across which heat is flowing

ΔT = the difference in temperature between two points a distance Δx apart

k = a proportionality constant characteristic of the material and is called the thermal conductivity of the material; its dimensions are W/m-°C.

Equation (5.1) as shown is valid only as long as the temperature varies linearly with x. This is not true in general for unsteady-state heat transfer or even steady-state heat transfer through a curved object, such as a cylinder or sphere. In such cases, calculus must be used, the temperature gradient being expressed in differential form.

Values of thermal conductivity k for various common materials are shown in Table 5.1. Study of Table 5.1 shows several interesting things. First, the thermal conductivity of copper is hundreds of times as great as that of other solids such as concrete, plaster, or wood, and more than 10,000 times as great as that of air. Second, thermal conductivity of metals decreases as the temperature increases, whereas the thermal conductivity of gases increases with increasing temperature. [Equation (5.1) can still be used if an average value of k is used for the temperature range of interest.]

As is well known, metals conduct electricity; nonmetals, such as concrete or wood, do not. Copper is an extremely efficient conductor of electricity. Heat is carried through electrically conductive solids by the electrons themselves, and this is a far more efficient transfer process than that of vibrational energy of atoms or molecules.

TABLE 5.1 Thermal conductivities of various materials

Material	Temperature (°C)	Thermal conductivity (W/m-°C)
Copper	20	386
Copper	600	353
Steel	20	54
Concrete	20	1.37
Plaster, gypsum	20	0.48
Wood (pine)	20	0.13
Air	0	0.023
Air	120	0.028

Convective Heat Transfer

A common heat transfer problem involves a flowing hot fluid in contact with the cold surface of a solid. (Or the surface might be hot and the fluid cold.) In such cases there is usually a slow-moving boundary layer of the fluid separating the faster-moving bulk of the fluid from the surface. The rate of heat transfer between the fluid and the surface depends on the thickness of this film or boundary layer, as well as the temperature difference across the boundary layer, and the thermal conductivity of the fluid in the boundary layer. Equation (5.2) is widely used for this type of problem

$$q' = hA\Delta T \qquad (5.2)$$

where:

q'	=	the rate of heat flow (watts)
A	=	the area across which heat is flowing
ΔT	=	the difference in the temperature of the bulk fluid and of the surface
h	=	the heat transfer coefficient

The essence of the problem is how to get a good value for h.

Since h is always proportional to k, the thermal conductivity of the fluid, it is convenient to incorporate h in a dimensionless group, hL/k, called the *Nusselt number* (Nu). L is a characteristic length, which must be defined for each application. Then, if Nu can be determined for a given case, and k and L are known, it is easy to calculate h, and then apply Eq. (5.2).

A large series of formulas for calculating Nu have been obtained from measurements and are available in heat transfer textbooks. These express the Nusselt number as a function of other dimensionless numbers, chiefly the

Reynolds number (Re), the Grashof number (Gr), and the *Prandtl number* (Pr). (The Prandtl number, involving the ratio of the fluid viscosity to its thermal conductivity, happens to have a nearly constant value of about 0.7 for air or combustion product gases, independent of temperature.)

For example, consider a hot fluid flowing past a cold cylinder of diameter L, the Reynolds number, based on L, being between 40 and 4000. Then, the following relation holds.

$$Nu = 0.68(Re)^{0.466}(Pr)^{0.33} \qquad (5.3)$$

As another example, consider a hot fluid flowing across a cold flat plate of length L, in the downstream direction. The boundary layer is initially laminar but becomes turbulent before the end of the plate is reached. For this case,

$$Nu = [0.037(Re)^{0.8} - 871](Pr)^{0.33} \qquad (5.4)$$

As a final example, consider a horizontal hot plate of diameter L, facing upward, with a buoyant upward flow of initially cold fluid resulting from contact with the hot plate. The formula for this case is

$$Nu = 0.14[(Gr)(Pr)]^{0.33} \qquad (5.5)$$

Equation (5.5) is valid only when the flow is turbulent, which is the case when [(Gr) (Pr)] is greater than about ten million. (For laminar conditions, another formula is available. [1])

Radiative Heat Transfer

The thermal radiation emitted by any object consists of electromagnetic radiations of various wavelengths or frequencies, all traveling at the velocity of light. Figure 5.1 shows the distribution of thermal radiative energy flux at various wavelengths. The figure shows that, at any temperature, the radiation has a peak intensity at some wavelength. It happens that the product of the temperature and the wavelength corresponding to the peak is a constant. The area under each curve in Figure 5.1 is proportional to the total radiative output. It is evident from Figure 5.1 that the radiative output increases substantially with increasing temperature.

The surface area A (m^3) of any body at any finite absolute temperature T (K) will emit thermal radiation q' (watts) in accord with the *Stefan–Boltzmann law:*

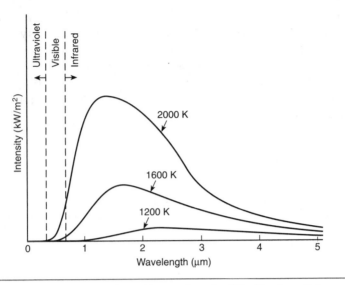

FIGURE 5.1 The variation with wavelength of blackbody radiation at various temperatures.

$$q' = 5.67 \cdot 10^{-8} A \varepsilon T^4 \tag{5.6}$$

where ε, a factor between zero and one, is the surface emissivity. A surface with emissivity one is referred to as a blackbody. There are a number of solids with emissivity greater than 0.9, for example, graphite, oxidized steel, oak, marble, gypsum, and any surface coated with flat black lacquer. For such materials, it may be convenient to set the emissivity equal to one in Eq. (5.6), with little loss of accuracy. On the other hand, there are many materials with low emissivities. Shiny metals will generally have emissivities below 0.1, and, in some cases, below 0.02. Bricks of various kinds may have emissivities from 0.4 to 0.8. In summary, the emissivity of the material involved must be known before Eq. (5.6) can be used correctly.

In a fire, the major source of radiant heat is the fire itself. More specifically, the soot particles in the flame, which give it a yellow or orange appearance, constitute a large source of radiation. Another (usually smaller, but not negligible) source of radiation is the energy radiated by certain hot gas molecules, primarily water vapor and carbon dioxide.

To calculate the emissivity of a gas–soot mixture, which is somewhat transparent to radiation, it is necessary to take into account the thickness of the mixture, as well as the concentrations of soot and emitting gases. If the thickness is great enough, the cloud is said to be optically thick, and the emissivity is unity. Soot concentration depends on the chemical nature of the combustible. (See Chapter 8, "Fire Characteristics: Gaseous Combustibles" and Chapter 11, "Combustion Products.")

When incoming radiation strikes a nontransparent surface, a fraction of it may be reflected and the remainder will be absorbed, heating the material. The fraction absorbed, α, is called the *absorptivity* of the material. The emissivity and the absorptivity of a material at a given temperature are the same. However, emissivity and absorptivity can vary somewhat with temperature. Even if the emitter and the receiver are of the same material, if the emitter is much hotter than the receiver the absorptivity may be different from the emissivity.

The calculation of radiative heat transfer between emitter and receiver of specified sizes, orientations, and separation distance often involves complex geometrical calculations. View factors are involved. These calculations have already been done for a number of cases, and the results are available as formulas or graphs.

Problems

1. Calculate the rate at which heat will flow through a square meter of a concrete wall 5 cm thick, if one side of the wall is at 700°C while the other side is at 20°C.

2. Air at 20°C is flowing transversely at 5 m/s across a metal cylinder 1 cm in diameter and 1 m long. The cylinder is at –15°C. Calculate the rate of heat transfer from the air to the cylinder.

3. A brick wall (emissivity 0.6) is at 400°C. Calculate the radiative flux emitted by 1 m^2 of this wall.

4. An orange flame of a fire is emitting radiation. What species within the flame are responsible for this radiation?

Reference

1. *The SFPE Handbook of Fire Protection Engineering*, 2nd ed., National Fire Protection Association, Quincy, MA, 1995. (See Chapter 1-2, "Conduction of Heat in Solids," by J. A. Rockett and J. A. Mike, pp. 1-25–1-38; Chapter 1-3, "Convection Heat Transfer," by A. Atreya, pp. 1-39–1-64; and Chapter 1-4, "Radiation Heat Transfer," by C. L. Tien, K. Y. Lee, and A. J. Stretton, pp. 1-65–1-79.)

PART II

Fire Protection Chemistry and Physics

LIBRARY
WAUKESHA COUNTY TECHNICAL COLLEGE
800 MAIN STREET
PEWAUKEE. WI 53072

WITHDRAWN

WITHDRAWN

CHAPTER 6

How Do Chemistry and Physics Relate to Fire Protection?

We must understand the nature of fire in order to know how to protect against it. Fire is one of the most dramatic examples of a chemical reaction, so an understanding of chemistry is essential in this task.

The materials of the world can be classified as follows:

1. Combustible gases, liquids, and solids
2. Noncombustible gases, liquids, and solids

Hydrogen, gasoline, and wood are, of course, combustible; nitrogen, water, and limestone are noncombustible. However, there are many materials or mixtures of materials that are not so readily classified as combustible or non-combustible. The chemistry of a material generally is the key to its degree of combustibility.

Initiation of fire requires not only a combustible material, but also oxygen, generally from the air, and usually an energy (heat) source to provide ignition (although certain materials are capable of spontaneous ignition). Three components — fuel, oxygen, and heat — are referred to as the *fire triangle. (See Figure 6.1.)* The chemistry of temperature-dependent oxidation reactions is crucial to a full understanding of the fire triangle.

A group of combustible materials will differ from one another in various ways relevant to fire protection:

1. Ease of ignition

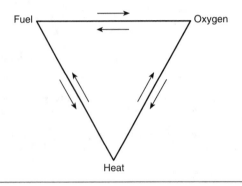

FIGURE 6.1 The fire triangle.

2. Rapidity of flame spread over exposed surfaces *(See Figure 6.2.)*

3. Peak heat-release rate per unit area of burning surface

4. Formation of protective char layer

5. Stoichiometric air requirement per gram of combustible

6. Output, per gram burned, of the following items:

   a. Heat

   b. Toxic gases

   c. Smoke

   d. Molten dripping material

7. Ease of extinguishment by the following methods:

   a. Water

   b. Water-based foam *(See Figure 6.3.)*

   c. Inert gases

   d. Special extinguishing agents

Furthermore, the properties of a given combustible material can be modified by the addition of fire-retardant chemicals or, for a synthetic material, by changes in its molecular structure.

Chemistry is essential to the understanding of differences in fire behavior of materials. Referring to the fire triangle (Figure 6.1), one can deduce that a fire can be extinguished in any of three ways:

1. Removing or blanketing the fuel

2. Cutting off or diluting the oxygen supply

FIGURE 6.2 Gas station fire.

FIGURE 6.3 Applying foam to cool and extinguish a bus fire.

3. Removing heat from the fire or the combustible, that is, cooling sufficiently to prevent the combustion reactions from continuing or to prevent the combustible from vaporizing *(See Figure 6.4.)*

There is also a fourth approach to fire suppression. It involves introducing chemicals that interrupt the flame chemistry of the combustion chain
reactions by removing free atoms and free radicals. For example, Halon 1301
(CF_3Br) can react rapidly with hydrogen (H) atoms in the flame to form HBr.
Furthermore, the HBr can react with other H atoms to form H_2 + Br, or it can
react with hydroxyl radicals (OH) to form H_2O + Br. Therefore, the net effect

FIGURE 6.4   Cooling and controlling a fire under a propane tank.

is to replace the reactive species H and OH in the flame by the less reactive species Br.

Water, which is the most commonly used extinguishing agent, can react chemically in unexpected ways with certain combustibles. For example, water might react with burning metals to generate hydrogen, or with glowing carbon to form carbon monoxide and hydrogen.

The various toxic gases that might be generated by a fire can cause entirely different, and sometimes opposing, physiological effects. For example, irritant gases such as acrolein or hydrogen chloride tend to cause a reduction in breathing rate whereas carbon dioxide stimulates breathing. The uptake rate of narcotic gases, such as carbon monoxide or hydrogen cyanide, depends, of course, on breathing rate. Thus, the question of what constitutes a dangerous concentration of toxic combustion gases is complex.

The ability to see through smoke in a corridor, as well as the ability of a smoke detector to give an early alarm, depends on the particle size as well as the concentration of the smoke. The smoke characteristics relate to the chemistry of the combustible and the combustion process.

Chemistry is a key factor in the scientific foundation of fire protection; however, two other key factors must be considered: heat transfer and fluid mechanics, both of which are considered to be physics. The radiative, convective, and conductive transfer of heat from the flame and the hot fire products to the burning and the not-yet-burning combustibles is vital in governing the growth of a fire. *(See Chapter 5.)* The buoyant rise of hot gases, drawing fresh air into the fire, the mixing of the fire gases with ambient air, and the motion of the fire products through a building, usually with layering, are important aspects of fires *(see Figure 6.5).* Heat transfer and fluid mechanics are interrelated because the motion of a hot gas affects the rate of heat transfer from the

FIGURE 6.5 Fire gases emerging from a building.

gas to the surroundings, and the temperature of a gas affects its buoyancy, which induces its motion.

The detailed study of how chemistry, fluid mechanics, and heat transfer interact to influence fire behavior is called fire dynamics. This text introduces fire dynamics; reference should be made to other books for more extensive information. [1, 2, 3]

References

1. Drysdale, D., *An Introduction to Fire Dynamics,* 2nd ed., J. Wiley, New York, 1998.
2. Quintiere, J. G., *Principles of Fire Behavior,* 1st ed., Delmar Publishers, Albany, NY, 1997.
3. *The SFPE Handbook of Fire Protection Engineering,* 2nd ed., National Fire Protection Association, Quincy, MA, 1995.

CHAPTER 7

The Combustion Process

WHAT IS COMBUSTION?

The term *combustion* usually refers to an exothermic (heat-producing) chemical reaction between some substance and oxygen. Chemical analysis of the combustion products would show the presence of certain molecules involving combinations of oxygen atoms with other types of atoms, such as CO_2, H_2O, SO_2, NO_2, Al_2O_3, or SiO_2.

A slow oxidative reaction is a reaction between some substance and oxygen that may require weeks or months to go to completion. Such a reaction (which is not combustion) releases heat so slowly that the temperature never increases more than a degree or so above the temperature of the surroundings. One example of this process is the rusting of a metal.

The difference between a slow oxidative reaction and a combustion reaction is that the latter occurs so rapidly that heat is generated faster than it is dissipated, causing a substantial temperature rise (at least hundreds of degrees and often several thousand degrees). Very often, the temperature is so high that visible light is emitted from the combustion reaction zone.

Combustion does not necessarily require oxygen molecules. For example, a mixture of potassium perchlorate ($KClO_4$) and polyethylene [$(C_2H_4)_n$], a rocket propellant, will undergo vigorous combustion in an inert atmosphere. Of course, $KClO_4$ contains oxygen atoms. However, other combinations of chemicals are completely lacking in oxygen but will burn with flames in the absence of air, for example:

1. $H_2 + Cl_2 \rightarrow 2\ HCl$.

2. (CF_3Br) [Halon 1301] $+ 2\ Mg \rightarrow {}^3/_2\ MgF_2 + {}^1/_2\ MgBr_2 + C_{(s)}$.

3. N_2H_4 [hydrazine] $\rightarrow N_2 + 2\ H_2$.

4. C_2H_2 [acetylene] $\rightarrow 2\ C_{(s)} + H_2$.

The concern in fire protection is generally with combustion reactions between various materials and the oxygen in the air, despite these counter-examples.

FLAMING AND NONFLAMING COMBUSTION

A flame is a gaseous oxidation reaction, which occurs in a region of space that is much hotter than its surroundings and generally emits light. Familiar examples include the yellow flame of a candle and the blue flame on a gas burner.

The flame has been referred to as gaseous. When a solid such as a match or a candle burns, a portion of the heat of the gaseous flame is transferred to the solid, causing the solid to vaporize. *(See Figure 7.1.)* This vaporization can occur with or without chemical decomposition of the molecules. If chemical decomposition occurs, it is called *pyrolysis*.

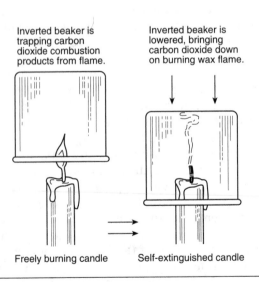

FIGURE 7.1 Flaming combustion (left) and self-extinguishment by its own combustion products (right) of a wax candle.

FIGURE 7.2 Examples of smoldering in two seat cushions. The cushion on the left burst into flames and then was extinguished; the cushion on the right — a new material being tested for use in furniture and mattress coverings — smoldered but did not burst into flames. (Courtesy of DuPont Industries.)

Another mode of combustion not involving any flame exists. It is called *smoldering, glowing,* or *nonflaming* combustion. A cigarette burns in this way. Upholstered furniture containing cotton batting or polyurethane foam can smolder. A large pile of wood chips, sawdust, or coal can smolder for weeks or even months.

Smoldering generally is limited to porous materials that can form a carbonaceous char when heated. As oxygen in the air slowly diffuses into the pores of the material, a glowing reaction zone occurs within the material, even though the glow might not be visible from outside. These porous materials are poor conductors of heat (i.e., good thermal insulators); therefore, even though the combustion reaction occurs slowly, enough heat is retained in the reaction zone to maintain the elevated temperature needed to sustain the reaction.

It is not uncommon for a piece of upholstered furniture, once ignited at a point, to smolder for several hours. *(See Figure 7.2.)* During this time, the reaction zone spreads only 5 or 10 cm from the ignition point, then suddenly the furniture can burst into flames. The rate of burning during flaming combustion is many times as great as during smoldering combustion.

HOW DOES COMBUSTION ORIGINATE?

Although oxygen is capable of reacting exothermically with many common materials (e.g., paper, wood, natural and synthetic fabrics, rubber, plastics, hydrocarbon liquids, and alcohol), the rate at which this reaction occurs is extremely slow at room temperature. However, if a high temperature can be generated in a small region where both combustible and oxygen are present, then the oxidation reaction will accelerate to a high rate in this region, generating its own heat and supply of free atoms and free radicals. The heat will spread to adjacent areas, and the combustion will spread; that is, the fire will grow.

The initial hot spot can be caused in various ways: by a static spark; by friction; by overheating in an electrical circuit; by a faulty space-heating device; by welding or cutting operations; by lightning; or by an already-burning object, such as a match, a cigarette, or a pilot flame. In some cases, self-heating, which leads to spontaneous ignition, is also possible. Details of the ignition process for gases, liquids, and solids are discussed in Chapters 8, 9, and 10, respectively.

HOW DOES COMBUSTION SPREAD?

The spread of combustion depends very much on whether the combustibles are gaseous, liquid, or solid. In the case of a flammable gas, it depends on the extent of gas and air mixing prior to combustion and on the degree of motion and turbulence of the gas. For a liquid, the spread of combustion depends on whether the combustion occurs in a still pool of liquid, a flowing liquid, a spray, a thin film, or a foam.

A solid might be in the form of a powder, a thin sheet (e.g., paper), or a thick solid object. (Combustion spreads much more rapidly over thin solids.) The spread rate on a vertical solid surface is much faster in the upward direction than in the sideways and downward directions. Fire spread over a horizontal surface depends on whether air currents are moving toward or away from the combustion zone.

The speed of fire spread can be as fast as several meters per second (about 10 ft/s) or as slow as 0.1 mm/s (0.004 in./s), depending on the conditions. However, in most cases of fire spread, the basic mechanism is the same: A portion of the heat produced by the existing fire will be transferred, either directly or indirectly, to a nearby combustible that is not yet burning, and this combustible will be heated to the temperature at which it can start burning. Of course, fire also can spread by melting and dripping of burning material or by airborne firebrands.

HOW IS COMBUSTION TERMINATED?

Combustion requires a high temperature, and the reactions must proceed fast enough at this high temperature to generate heat as fast as it is dissipated so that the reaction zone will not cool down. If anything is done to upset this heat balance, such as introducing a coolant, then it is possible that the combustion will be extinguished. It is not necessary for the coolant to remove heat as fast as it is being generated, because the combustion zone in a fire is already losing some heat to the cooler surroundings. In some cases, only a modest additional loss of heat is needed to tip the balance toward extinguishment.

Extinguishment can be accomplished by cooling either the gaseous combustion zone or the solid or liquid combustible. In the latter case, the cooling prevents the production of combustible vapors. (This is probably the primary mode of action when a wood fire is extinguished by applying water.)

As a simple extinguishment experiment, support a stiff vertical strip of paper from the bottom, as shown in Figure 7.3, and attach a paper clip 5 cm (2 in.) from the top. Ignite the top of the paper with a match. Observe the burning zone as it moves downward. When the fire reaches the paper clip, it is extinguished. The metal of the paper clip acts as a "heat sink" and upsets the balance between heat production and heat loss. The flame cannot heat the paper to its pyrolysis temperature in the available time because of heat absorption by the paper clip.

Extinguishment also can be accomplished by introducing a barrier between the combustible surface and the flame, for example, by providing a blanket of aqueous foam. *(See Figure 7.4.)* The foam not only cools the surface but also prevents the fire's radiant heat from reaching the surface and promoting vaporization.

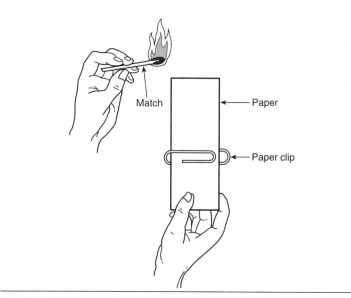

FIGURE 7.3 Demonstration of fire extinguishment by removal of heat.

FIGURE 7.4   Fire extinguishment with foam.

Cutting off the air supply also will cause extinguishment. *(See Figure 7.1.)* Alternatively, the oxygen content of the air supply can be reduced by diluting the air with carbon dioxide, nitrogen, steam, or combustion products. In some cases, it might be possible to separate the combustible from the fire; for example, a burning object such as a mattress could be removed from the fire compartment, or explosives could be used to blow the flame away from burning objects (such as in fighting forest fires).

Gaseous flames contain free radicals that participate in chain reactions. Some bromine-containing chemicals (e.g., CF_3Br) reduce the concentration of free radicals when added to flames and slow down the combustion reactions. Combined with heat loss to the cooler surroundings, this reduction in combustion-reaction rate can lead to extinguishment. Unfortunately, such agents are expensive, produce toxic and corrosive decomposition products, and recently have been found to have an adverse effect on the earth's protective ozone layer.

The foregoing review of combustion is abbreviated. For more detail, refer to texts by Drysdale [1], Glassman [2], and Strehlow [3].

Problems

1. When is oxidation *not combustion? When is combustion not oxidation?*

2. What kinds of materials can undergo smoldering combustion?

3. Name six ways in which a fire can originate.

4. What is the basic mechanism of fire spread?

5. What are the three fundamental ways of extinguishing a fire?

6. Perform the experiment shown in Figure 7.3.

References

1. Drysdale, D., *An Introduction to Fire Dynamics*, 2nd ed., J. Wiley, New York, 1998.

2. Glassman, I., *Combustion*, 3rd ed., Academic Press, New York, 1996.

3. Strehlow, R. A., *Combustion Fundamentals*, 1st ed., McGraw-Hill, New York, 1984.

CHAPTER 8

Fire Characteristics: Gaseous Combustibles

TYPES OF GASEOUS FLAMES

Flames can be categorized as *premixed flames* or *diffusion flames*, that is, fuel gas mixed with oxygen before or during combustion, respectively. In addition, they can be categorized as *laminar* or *turbulent* flames, as well as *stationary* or *propagating* flames. Any combination is possible, such as a turbulent propagating premixed flame or a laminar stationary diffusion flame. Each of these categories, as well as the possibility that a flame could be a detonation, is discussed.

Premixed Versus Diffusion Flames

Imagine a compartment containing 9.5 percent methane gas (CH_4) by volume and 90.5 percent air, thoroughly mixed. Because air contains 21 percent oxygen by volume, and because 21 percent of 90.5 is 19, the compartment must contain 19 percent oxygen by volume. Recall from Chapter 3 that x volume percent of a gas is the same as x mole percent. Therefore, the ratio of the moles of oxygen in the compartment to the moles of methane is 19/9.5, or 2. This is a stoichiometric mixture, according to the equation

$$CH_4 + 2\ O_2 \rightarrow CO_2 + 2\ H_2O$$

If an ignition source, such as a spark, is provided in the center of the compartment, then a small blue spherical flame will form around the spark and spread radially outward at about 3 m/s (10 ft/s).

Similar behavior would result if the methane percentage in air is somewhat lower or somewhat higher than 9.5 percent, except that the flame would propagate more slowly. If less than 5 percent or more than 15 percent of methane is present, the mixture would be too far from stoichiometric, and no ignition would occur. For combustible mixtures with greater than 9.5 percent methane ("rich" mixtures), there would be insufficient oxygen to completely oxidize the CH_4 to CO_2 and H_2O, and the products would include some CO, some H_2, and, for very "rich" mixtures, some solid carbon (*soot*).

This type of flame, whether stoichiometric or not, is a *premixed flame*. Premixed flames can be either moving through space (propagating), as described above, or *stationary*, as will be described later.

The contrasting type of flame is a *diffusion flame*. Assume that there is a cloud of methane, resulting from a sudden release of gas from a tank, surrounded by air, but mixing has not occurred yet except in a thin zone at the interface between methane and air. If an ignition source is provided at this interface, then combustion will spread rapidly over the surface of the cloud. Subsequently, the bulk of the methane within the cloud will burn more slowly, as air and methane interdiffuse. Meanwhile, the hot burning ball of gas will rise. The flame will be yellow.

Diffusion flames can be stationary. A familiar example is a candle flame, in which molten wax evaporates from the wick and the wax vapor interdiffuses with oxygen from the surrounding air. Most fires burn in this way, except on a larger scale.

In summary, premixed flames burn more rapidly than diffusion flames, and the chemistry is different (i.e., blue flames versus yellow flames. (See Chemical Mechanisms of Combustion later in this chapter for more details.)

Laminar Versus Turbulent Flames

Large flames or flames burning in high-velocity flows generally are turbulent. That is, the velocity and temperature fluctuate at any selected point in the flame, and the track of any particle moving through the flame (a streamline) is erratic, with many changes of direction, rather than a straight or gently curving line.

On the other hand, small flames such as candle flames or the flame cones on a domestic gas burner generally are laminar; that is, the streamlines are smooth and the fluctuations are absent or negligibly small.

The presence of turbulence in a flame enhances heat transfer and mixing and can even affect the chemistry. Accordingly, rates of combustion are considerably higher in turbulent than in laminar flames. This phenomenon makes it difficult to predict large-scale fire behavior from small-scale (bench-top) fire tests.

As a general rule, a diffusion flame taller than 0.3 m (1 ft) will be turbulent, and a diffusion flame shorter than 0.1 m (4 in.) will be laminar unless a high-velocity jet is involved.

Stationary Versus Propagating Flames

A premixed flame can propagate through a combustible gas mixture at 1 m/s (3 ft/s) or faster. However, there are several techniques for stabilizing a premixed flame so that it burns in a fixed position. Figure 8.1 shows three burner arrangements for studying premixed flames.

Such burners permit combustion scientists to measure flame properties accurately. For example, the gas-flow rate at which the flame will stabilize is a good way to measure the speed of flame propagation. The distribution of temperature and chemical species within the flame can be measured. These measurements lead to a detailed understanding of the combustion process.

Figure 8.2 shows four arrangements for stabilizing a diffusion flame. Once again, measurements of the stabilized flame structure lead to understanding of this type of flame, which more nearly resembles a fire.

IGNITION OF GASES

Whether or not a mixture of a given combustible gas plus oxygen plus inert gas can be ignited depends on its composition, temperature, and pressure. Assuming that the composition, temperature, and pressure are such that ignition is possible, then the ease of ignition can be considered. It will be different for each gas.

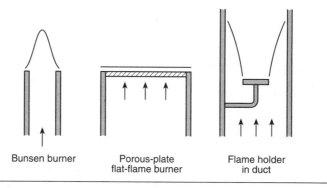

| Bunsen burner | Porous-plate flat-flame burner | Flame holder in duct |

FIGURE 8.1 Techniques for stabilizing premixed flames. The arrows represent the flow of a fuel–air mixture.

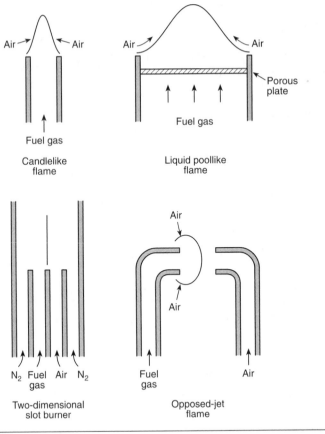

FIGURE 8.2 Techniques for stabilizing gaseous diffusion flames.

If ignition is accomplished by a spark that jumps across a short gap produced by discharging a capacitor, it is possible to measure the minimum energy of the weakest spark that will cause ignition. Such experiments show that a near-stoichiometric mixture of a combustible gas and air requires much less spark energy than does a mixture with substantial excess air (a "lean" mixture) or a mixture with substantial excess combustible (a "rich" mixture).

Table 8.1 shows values of minimum ignition energies for a series of combustible gases and vapors mixed either with air or with pure oxygen. In either case, the mixture composition is near-stoichiometric or near-optimum for ignition. Table 8.1 shows the following:

TABLE 8.1 Minimum spark ignition energies of combustible gases and vapors in air or oxygen at 25°C and 1 atm

Combustible	Minimum ignition energy (µJ)	
	Air	Oxygen
Methane, CH_4	300	3
Propane, C_3H_8	260	2
n-Hexane, C_6H_{14}	290	6
Ethane, C_2H_6	260	2
Ethylene, C_2H_4	70	1
Acetylene, C_2H_2	17	0.2
Carbon disulfide, CS_2	15	—
Hydrogen, H_2	17	1.2

Source: Kuchta, p. 33. [1]

1. Combustible-oxygen mixtures are far easier to ignite with weak sparks than combustible-air mixtures.

2. The family of saturated hydrocarbons (C_nH_{2n+2}) all ignite similarly in air.

3. Progressive unsaturation (double or triple bonds) in a hydrocarbon molecule favors much easier ignition, as in the sequence C_2H_6, C_2H_4, C_2H_2.

4. Certain gases — carbon disulfide, hydrogen, and acetylene — can be ignited with sparks less than one-tenth as strong as those required to ignite methane, propane, or n-hexane.

A gas mixture also can be ignited by contact with a hot solid surface. The minimum surface temperature at which ignition occurs can be measured for any combustible mixture, but unfortunately the result depends on the size, shape, orientation, and nature of the surface as well as the state of motion of the gas. For example, Jost [2] reported that seventeen different investigations of the ignition temperature of hydrogen–air mixtures gave results varying from a low of 410°C to a high of 930°C. Mullins [3] reported ethyl ether–air ignition temperatures from three sources as 186°C, 343°C, and 491°C. Accordingly, any reported value of gas ignition temperature is valid for only one set of conditions. However, if any one experimental technique is used consistently, *relative* ignition temperatures for a series of gases can be obtained. With this caution in mind, some ignition temperatures of gases in air are presented in Table 8.2.

TABLE 8.2 Thermal ignition temperatures of selected gases and vapors in air

Gas or vapor	Thermal ignition temperature (°C)
Methane, CH_4	537
Propane, C_3H_8	450
Hexane, C_6H_{14}	225
Octane, C_8H_{18}	206
Ethane, C_2H_6	472
Ethylene, C_2H_4	450
Acetylene, C_2H_2	305
Carbon disulfide, CS_2	90
Ethyl ether, $C_2H_5OC_2H_5$	160
Hydrogen, H_2	400

Source: Fire Protection Handbook. [4]

For the same group of gases, Table 8.2, thermal ignition, shows some differences in trends from Table 8.1, spark ignition. First, the thermal ignition data show that carbon disulfide is far easier to ignite than hydrogen, while the spark ignition data show that almost the same ignition energies are required for this pair. Second, the thermal ignition data show progressive decrease of ignition temperature with increasing molecular weight for the saturated hydrocarbons, while the spark ignition data show no such trend. Both Tables 8.1 and 8.2, however, show that acetylene is easier to ignite than ethane. Ethyl ether is included in Table 8.2 because of its remarkably low ignition temperature.

The ignition of a gas mixture is a surprisingly complex process, but it can be accomplished with a relatively weak spark or with a moderately hot surface. Note that carbon disulfide vapor, with an ignition temperature of 90°C, can be ignited by a steam pipe at 100°C.

FLAMMABILITY LIMITS AND PROPAGATION RATES OF PREMIXED FLAMES

There has been some confusion about the words *flammable, nonflammable,* and *inflammable.* A Merriam-Webster® dictionary defines inflammable as "flammable"; to become inflamed is to catch on fire. On the other hand, the words inaccurate, insane, inconceivable, and so on, mean the opposite of accurate, sane, conceivable, and so on. Therefore, if a container is labeled "inflammable," one might be uncertain as to whether it meant flammable or nonflammable. This potential ambiguity is unacceptable from a fire safety standpoint. Consequently, the fire protection profession strongly urges that the words flammable and nonflammable be used as appropriate, and that the words inflammable and inflammability never be used.

Suppose there is a stoichiometric mixture of a combustible (or flammable) gas, such as methane and pure oxygen, at 25°C. If ignited, this mixture will burn vigorously, with a flame temperature of 2760°C. If, instead, there is a stoichiometric mixture of methane and air (9.5 percent CH_4, 19 percent O_2, and 71.5 percent N_2), it also will burn but not quite as vigorously, and the flame temperature will be 1940°C. The flame reactions will proceed more slowly because of the lower temperature. The reason for this lower temperature is the presence of inert nitrogen, which absorbs some of the heat released by the $CH_4 + 2\ O_2$ reaction.

If still more nitrogen is added to the methane–air mixture, it will reach a point where the mixture is no longer flammable. The mixture of borderline flammability will have 6.2 percent CH_4, 12.4 percent O_2, and 81.4 percent N_2, and will burn with a flame temperature of about 1200°C. The explanation has been offered that the hydrocarbon oxidation reactions in a flame do not occur rapidly enough below about 1200°C to overcome either the heat losses from the flame to the surroundings or the rate of loss of free atoms and radicals in the flame by recombination. Another explanation for flammability limits is that highly diluted flames propagate so slowly that free convective motions of the flame become larger than the propagation speed, so the flame structure is disrupted. Whatever the reason, highly diluted mixtures are not flammable.

Assume that a stoichiometric methane–air mixture exists that contains 2 molecules of oxygen for each molecule of methane. If either additional air (to form a "lean" mixture) or additional methane (to form a "rich" mixture) is added, the added material will act to reduce the flame temperature and, if enough is added, the mixture will be rendered nonflammable.

To better understand why the flame temperature is reduced, consider an extreme case: 1 percent methane and 99 percent air. The CH_4 molecules can each react with 2 O_2 molecules, so 1 + 2 (or 3 percent) of the molecules can react with each other, generating heat, while the other 97 percent of the molecules contribute no energy but absorb energy from the reaction. Accordingly, the temperature rises only slightly, and comes nowhere near 1200°C, which is the approximate temperature needed to sustain the reaction.

Figure 8.3 shows the *flammability limits* of methane–air mixtures, as well as the effect of introducing additional nitrogen. First, look at the ordinate methane, volume percent. The methane–air flammability range is from 5 percent (lower limit, or "lean" limit) to 15 percent (upper limit, or "rich" limit). The stoichiometric composition of 9.5 percent is approximately in the middle.

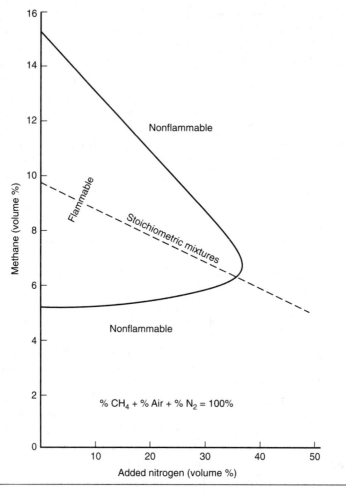

FIGURE 8.3 Limits of flammability of various methane–air–nitrogen mixtures at 25°C and 1 atm. Source: Zabetakis, p. 29. [5]

Next, if more than about 35 percent additional nitrogen is added to a stoichiometric methane–air mixture, a boundary is crossed into a region of nonflammable mixtures. Furthermore, Figure 8.3 shows that somewhat lesser percentages of additional nitrogen could be added to "lean" or "rich" mixtures, rendering them nonflammable.

Measurements have been made of limits of flammability for hundreds of gases and vapors. (A *vapor* is simply a gas that is readily condensable to a liquid; all gases are condensable if cooled sufficiently.) Zabetakis [5] made an extensive compilation of flammability limits, which also is reproduced in a book by Glassman. [6] Table 8.3 presents some selected values from this compilation.

TABLE 8.3 Limits of flammability of gases and vapors in air at 25°C and 1 atm

	Flammability limits (volume %)	
Combustible	Lower limit	Upper limit
Acetone	2.6	13
Acetylene	2.5	100
Ammonia	15	28
Carbon disulfide	1.3	50
Ethane	3.0	12.4
Ethanol (ethyl alcohol)	3.3	*
Ethylene	2.7	36
Ethyl ether	1.9	36
n-Hexane	1.2	7.4
Hydrogen	4.0	75
Methane	5.0	15.0
Methanol (methyl alcohol)	6.7	*
Propane	2.1	9.5

Source: Zabetakis, pp. 114–116. [5]

*The vapor pressure of these liquids at 25°C is insufficient to reach the upper limit concentration.

Note that in Table 8.3 some substances have extremely wide limits of flammability and are, therefore, especially hazardous (e.g., hydrogen, 4 to 75 percent; carbon disulfide, 1.3 to 50 percent; acetylene, 2.5 to 100 percent). Most hydrocarbon gases have narrower limits (e.g., *n*-hexane, 1.2 to 7.4 percent). The reason that acetylene has an upper limit of 100 percent is that pure acetylene, in the absence of oxygen, can burn with a flame that produces hot carbon and hydrogen as products.

The foregoing flammability limits were measured at 25°C. If the temperature is higher than this, the flammability limits are wider.

Burning Velocity

Any mixture within the flammability boundaries is capable of propagating a flame. Theoretically, a flame is capable of moving through the gas mixture at a speed that depends only on the intrinsic properties of the gas (i.e., composition, temperature, pressure). This theoretical flame speed often is called the *burning velocity.* There are three reasons why a flame can actually move ·considerably faster than the burning velocity.

1. If the flame raises the temperature of the gas from, say, 300 K to 2100 K, this causes a sevenfold expansion of the hot portion of the gas, according to the perfect gas law. The expansion causes motion of the gas and carries the flame toward the unburned gas.

2. The concept of burning velocity assumes that the gas is moving into the flame surface, or the flame surface is moving into the gas, so that the velocity vector is normal (perpendicular) to the flame surface. Very often, this velocity vector is at some angle θ to the flame; that is, the flame is inclined to the flow. In such a case, the burning velocity is the normal component of the velocity vector V, and is equal to $V \sin \theta$.

3. The burning velocity is based on laminar flame propagation. A turbulent premixed flame will propagate several times as fast as a laminar one.

With these facts in mind, look at Figure 8.4, which shows laminar burning velocities versus composition for various combustible-air mixtures. Hydrogen and acetylene have exceptionally high burning velocities. Most combustible-air mixtures involving compounds without double or triple bonds have peak burning velocities of about 40 to 50 cm/s. For more information, Glassman [6] lists burning velocities of about 100 compounds.

If the combustible mixture is at an elevated temperature before ignition, the burning velocity will be greater. If the oxidant is pure oxygen instead of air, the burning velocity will be 5 to 10 times as great. The burning velocity also will depend somewhat on the ambient pressure.

Combustion scientists now understand in considerable detail the relation between the burning velocity of a mixture and the rates of the chemical oxidation reactions occurring in its flame. Refer to Glassman [6], Drysdale [7], and Strehlow [8] for additional information. To summarize a principal finding, the burning velocity is approximately proportional to the square root of the product of the mean rate of heat generation in the flame (per unit volume of reaction zone) and the thermal conductivity of the gas mixture.

Deflagration Versus Detonation

A premixed flame may burn either as a *deflagration* or as a *detonation*. The descriptions up to now have been of deflagrations. The differences are (1) a deflagration propagates at subsonic velocity; a detonation propagates at supersonic velocity, relative to the unburned gas; and (2) an element of gas passing through a deflagration experiences a small decrease in pressure and a large decrease in density; an element passing through a detonation experiences a large increase in pressure and a moderate increase in density.

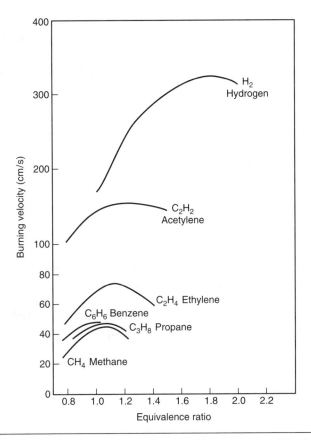

FIGURE 8.4 Laminar burning velocities of some combustibles in air at 25°C and 1 atm. Note that the equivalence ratio of a mixture is its fuel–air ratio divided by the stoichiometric fuel–air ratio. Source: Glassman. [6]

The speed with which a deflagration moves (i.e., the burning velocity) depends on the chemically controlled rate of heat release in the flame. The speed of a detonation is independent of the rates of the heat-generating chemical reactions. The reason for this is that a detonation is a shock wave that, by compression, heats a reactive gas mixture to a temperature high enough that the heat of combustion is released. It is possible to calculate the velocity of detonation rather precisely from a knowledge of the heat of combustion and the physical properties of the mixture. Of course, there are marginal conditions, such as small tube diameter or highly diluted mixture composition at which detonations no longer can occur, and these conditions are influenced by chemical reaction rates.

The damage caused by a detonation is usually due to the very high pressure generated. For example, a detonating methane–oxygen mixture, originally at 1 atm, generates a pressure of over 30 atm, and moves at 2500 m/s. (Detonations of liquid or solid explosives produce pressures of tens of thousands of atm.)

In general, mixtures of combustible gases with oxygen are much more likely to detonate than mixtures with air. However, mixtures with air can detonate, under suitably confined conditions, or with a sufficiently powerful initiating event. For example, ignition of a gas–air mixture at the closed end of a long tube can produce an accelerating flame, becoming turbulent, with a buildup of pressure and a transition to detonation.

Further information on detonation is given by Strehlow. [8]

CHEMICAL MECHANISMS OF COMBUSTION

Much is known about the chemical reactions occurring in both premixed flames and diffusion flames. When hydrogen is the combustible, the reactions are similar in premixed and diffusion flames. However, when the combustible molecules contain the element carbon, as is usually the case, then yellow soot is seen in diffusion flames, whereas premixed flames are not yellow unless they are extremely fuel-rich. The yellow soot consists of tiny hot particles (10^{-6} to 10^{-5} cm diameter) of which carbon is the main constituent. The empirical formula of the soot is approximately $(C_8H)_n$.

Soot is important in fires for two reasons:

1. It affects the radiative heat transfer rate from flames and therefore the combustion rate.

2. It is a principal source of the vision-obscuring smoke produced in fires.

The soot in diffusion flames is formed when the combustible molecules become heated and decompose as they move toward the oxidant side of the flame, but have not yet reached oxidizing molecules. Later, when the soot encounters oxygen-containing molecules (especially the free radical OH), part or all of the soot particles can be oxidized to CO or CO_2. Clearly this type of chemistry in a diffusion flame has no parallel in a premixed flame where substantial amounts of oxygen are present originally with the combustible.

The soot particles can radiate more effectively than the gas molecules, and at least 80 percent of the thermal radiation from a luminous (sooty) flame comes from the soot, although 10 to 20 percent comes from the hot CO_2 and H_2O molecules. Accordingly, a diffusion flame will radiate more than a nonsooty premixed flame, causing the diffusion flame to burn at a lower temperature because some of the combustion heat is lost. Because of this lower temperature, the oxidation reactions in a diffusion flame can cease before completion under some conditions, and substantial carbon monoxide, unconsumed soot, and other unoxidized molecules can be found in the products.

(Of course, incomplete combustion also will occur when the available air is insufficient.)

Now consider the specific sequences of chemical reactions occurring, first in hydrogen flames and then in methane flames.

The H_2–O_2 reaction is well understood. The reaction does not occur by H_2 and O_2 molecules colliding with each other, but by a chain reaction involving the free atoms and radicals H, O, and OH. The most critical reaction is

$$H + O_2 \rightarrow OH + O$$

In this reaction, a single H atom (produced by the ignition source) can react with a stable molecule, O_2, producing two highly reactive species, OH and O. The OH (hydroxyl radical) can react very rapidly with H_2:

$$OH + H_2 \rightarrow H_2O + H$$

It is consumed thereby, but produces an H atom that can continue the chain reaction. Meanwhile, the O atom produced in the first reaction can react very rapidly with H_2 to form two additional chain carriers:

$$O + H_2 \rightarrow OH + H$$

Thus, as shown in Figure 8.5, a single H atom, when introduced into an H_2–O_2 mixture at an elevated temperature, will be transformed by a sequence of rapid reactions (requiring a fraction of a millisecond) to form two molecules of H_2O and three new H atoms. Each of these new H atoms can immediately initiate the same sequence, and a *branching chain reaction* is produced; it continues until the reactants are consumed. Then, the remaining H, O, and OH species recombine according to the reactions

$$H + O \rightarrow OH$$

and

$$H + OH \rightarrow H_2O$$

In the methane–oxygen reaction, the reaction step

$$H + O_2 \rightarrow OH + O$$

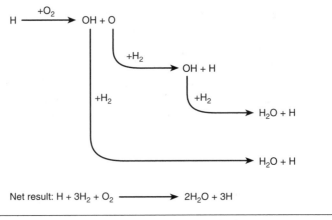

Net result: $H + 3H_2 + O_2 \longrightarrow 2H_2O + 3H$

FIGURE 8.5 Chain reaction mechanism in the hydrogen–oxygen flame.

also is important, but the sequence of reactions is considerably more complex. About 100 individual reaction steps are believed to be important, and the rates of most of them are known. Figure 8.6 shows some of the most important steps in the methane–oxygen premixed flame. (Nitrogen from the air, if present, can be considered to be inert, unless there is interest in traces of nitrogen oxides, which can form.)

Note in Figure 8.6 that there are several parallel paths by which the reaction proceeds, and also that the stable intermediate CO must form before the formation of CO_2. If the flame gases cool before the CO is completely converted to CO_2 via oxidation by OH, then CO will appear in the products, even if an excess of oxygen is available in the environment.

In a methane–air diffusion flame, the reactions would be similar to the foregoing description except on the side of the flame toward the methane. There, the presence of heat and the near-absence of oxygen cause decomposition (pyrolysis) reactions to occur. The exact sequence of chemical steps by which methane and other hydrocarbons are converted partially to soot has not been established fully; however, research suggests sequences of reactions similar to those shown in Figure 8.7 (*Twenty-first Symposium (International) on Combustion [9], Twenty-second Symposium (International) on Combustion [10]*).

A series of reactions occurs, starting with CH_4, having 1 carbon atom per molecule, in which progressively larger molecules form, with more carbon atoms and lower hydrogen-to-carbon ratios. When the molecules become large enough (usually 6 carbon atoms), they can become *cyclic* (form closed rings). Soot contains polycyclic *aromatic hydrocarbons*, some examples of which are shown in Figure 8.8.

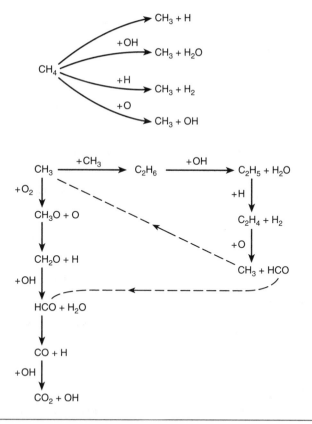

FIGURE 8.6 Some reaction steps in the methane–oxygen premixed flame in addition to the H, O, OH, H_2, O_2 reactions of Figure 8.5.

If, instead of methane, the combustible had been a compound with more than 1 carbon atom, the sequence of reactions building up to soot would be shorter, so that propane, C_3H_8, for example, would be expected to burn with a sootier flame than methane. This is indeed the case. Also, because carbon compounds with a double bond (C_nH_{2n}) or two double bonds (C_nH_{2n-2}) or a triple bond (also C_nH_{2n-2}) are closer to the end product ($C_8H)_n$ than are saturated hydrocarbons (C_nH_{2n+2}), they should be expected to give sootier flames. This also is the case. Finally, because soot has a cyclic structure, it would be expected that if the original fuel had a cyclic structure, for example, benzene (C_6H_6) or toluene ($C_6H_5CH_3$), this would cause an extremely sooty flame, and it does.

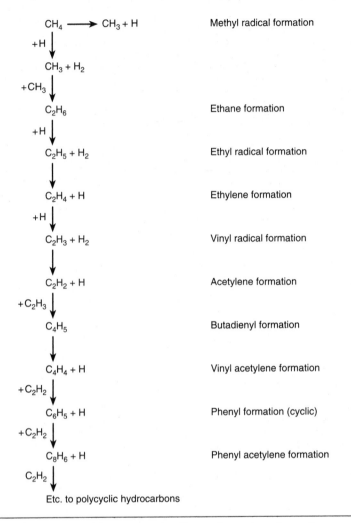

$CH_4 \longrightarrow CH_3 + H$	Methyl radical formation
$+H \downarrow$	
$CH_3 + H_2$	
$+CH_3 \downarrow$	
C_2H_6	Ethane formation
$+H \downarrow$	
$C_2H_5 + H_2$	Ethyl radical formation
$\downarrow$	
$C_2H_4 + H$	Ethylene formation
$+H \downarrow$	
$C_2H_3 + H_2$	Vinyl radical formation
$\downarrow$	
$C_2H_2 + H$	Acetylene formation
$+C_2H_3 \downarrow$	
C_4H_5	Butadienyl formation
$\downarrow$	
$C_4H_4 + H$	Vinyl acetylene formation
$+C_2H_2 \downarrow$	
$C_6H_5 + H$	Phenyl formation (cyclic)
$+C_2H_2 \downarrow$	
$C_8H_6 + H$	Phenyl acetylene formation
$C_2H_2 \downarrow$	
Etc. to polycyclic hydrocarbons	

FIGURE 8.7 Possible reactions leading to formation of soot from methane.

Table 8.4 shows the height of the shortest laminar diffusion flame, for a series of combustibles, that will release black smoke from its tip. The large differences caused by chemical variations are apparent. This *smoke-point height* is a convenient and useful measure of combustible smokiness and of flame radiative emission.

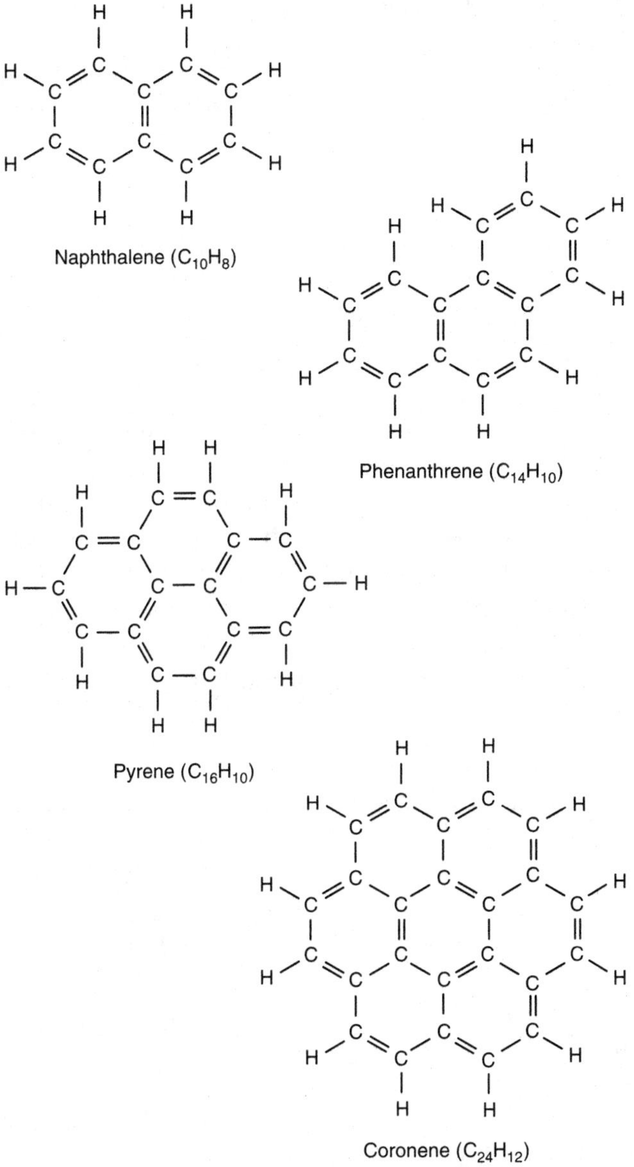

FIGURE 8.8 Four soot precursor molecules: polycyclic aromatic hydrocarbons.

RADIATION FROM FLAMES

As previously noted, radiation is primarily from soot particles in luminous flames and secondarily from CO_2 and H_2O molecules. (N_2 and O_2 molecules do not radiate.) The radiation emitted from flames is important for at least three reasons.

1. The energy feedback from the flame to the burning material is often primarily by *radiation*, rather than *convection* (flow of hot gases) or conduction (as through a wall). Accordingly, the flame radiation strongly influences the rate of burning.

2. The *spread* of a fire to nearby combustibles is often by radiative transfer. For example, a radiative flux of 35 kW/m^2 impinging on vertical particleboard (wood) will cause ignition in about 50 seconds. The more intense the radiation, the more rapid the fire spread.

3. The radiation from a sizable fire can be so intense that fire fighters might not be able to approach the fire without protection. The threshold of pain for incident radiation on human skin is about 5 kW/m^2 for 10 seconds. [13]

In view of the importance of flame radiation, it is appropriate to consider the factors involved in predicting the intensity of radiation from a flame. The radiation impinging on a target will depend on the following:

Table 8.4 Laminar smoke-point height for various combustibles

Combustible	Smoke-point height (cm)
Methane, CH_4	Flame too tall; becomes turbulent
Ethane, C_2H_6	24.3
Propane, C_3H_8	16.2
n-Butane, n-C_4H_{10}	16.0
Ethylene, C_2H_4	10.6
Propylene, C_3H_6	2.9
Isobutylene, i-C_4H_8	1.9
Acetylene, C_2H_2	1.9
1,3-Butadiene, 1,3-C_4H_6	1.5
Benzene (vapor), C_6H_6	0.8*
Toluene (vapor), $C_6H_5CH_3$	0.6*
Naphthalene (vapor), $C_{10}H_8$	0.4*

Source: Schug et al. [11], except data marked (*), which are from Hunt, p. 603 [12].

1. Surface area of the flame facing the target
2. Flame temperature
3. Flame emissivity (which depends on its sootiness and its thickness)
4. Distance from the flame to the target.

If there are cooler smoke or fog droplets intervening between the flame and the target, this interference will scatter the radiation and reduce the intensity on the target.

If the rate of burning is known, it is possible to estimate the intensity of the flame radiation by the following procedure: Assume that a fire is consuming a combustible at the rate of 100 g/s, and the heat of combustion of the material is 50 kJ/g. Then, the theoretical rate of energy release is $100 \cdot 50$ = 5000 kJ/s, or 5000 kW. Assume that 30 percent of this energy is radiated from the flame, while 65 percent is convected upward with the fire products and 5 percent is not released because of incomplete combustion. It follows that 30 percent of 5000 kW, or 1500 kW, are radiated by the flame. This radiation goes in all directions.

Suppose one needs to estimate the radiative flux 10 m (33 ft) from the center of the flame. A hypothetical sphere centered on the flame has a surface area of

$$4\pi r^2 = 4(3.14)(10)^2 = 1256 \text{ m}^2$$

If the radiation is distributed equally in all directions (isotropic), then the fraction of the 1500 kW that impinges on 1 m^2 of the target is $^1/_{1256}$. Therefore, the target is receiving 1500/1256 = 1.2 kW/m^2. This radiation intensity would be far from sufficient to ignite a combustible material, but the radiation would be more intense closer to the flame.

More refined methods of calculation exist, taking into account the flame shape, effect of intervening smoke, and so on, but these methods are not discussed here. The role of chemistry in this process is in influencing the fraction of combustion energy appearing as radiation (30 percent in the preceding example).

Table 8.5 shows that this fraction of combustion energy is very dependent on the chemical nature of the combustible, ranging from 9 percent (hydrogen) to 43 percent (1,3-butadiene). Table 8.5 also shows that those flames that radiate intensively are characterized by incomplete combustion. The reason is that the loss of heat by radiation causes a lower temperature in the flame, which slows down the chemical oxidation reactions.

Smoke-point heights for various combustibles were discussed in the previous section. It is interesting to note the correlation between the smoke-point heights of laminar flames and the radiative fraction for turbulent flames for the same combustible. (See Figure 8.9.)

TABLE 8.5 Fraction of combustion energy radiated from various diffusion flames of gases in air

Gas	Percent of theoretical combustion energy		
	Radiated	Convected	Not Released
Hydrogen	9[a]	91	0[d]
Methane	18[b]	81	1[d]
Ethane	20[c]	79	1[c]
Propane	27[c]	68	5[c]
Ethylene	32[c]	59	9[c]
Propylene	39[c]	50	11[c]
1,3-Butadiene	43[b]	37–42	15–20[d]

[a]Estimate, based on Fishburne and Pergament, 1979, p. 1069. [14]
[b]Markstein, 1985, p. 1057. [15]
[c]Tewarson, pp. 3-35–3-124. [16]
[d]Author's estimate.

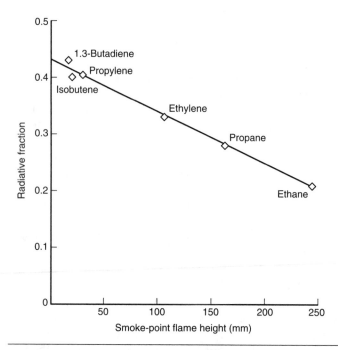

FIGURE 8.9 Radiative fraction for turbulent fuel-jet flames of various hydrocarbon fuels versus smoke-point laminar flame height. Source: Markstein. [15]

The hot smoke layer under the ceiling of a room in which there is a fire also is an important source of radiation, as well as the ceiling itself, once it becomes hot. These radiation sources are never as hot as the flame, but usually they extend over much more area, so that the radiant flux from them is often important in promoting flame spread. Figure 8.10 shows the radiative flux emitted by a hot surface at various temperatures.

INFORMATION ABOUT SPECIFIC HAZARDOUS GASES

Hydrogen (H₂)

Hydrogen is odorless and burns with a flame that is almost invisible (except in a darkened room). Its mixtures with air are ignited very easily with a low-energy spark, and the limits of flammability are unusually wide. *(See Figure 8.11.)* The burning velocity of hydrogen is higher than for any other combustible. Escaped hydrogen, being much lighter than air, will rise and form a layer under a ceiling. A hydrogen–air mixture in a suitably confined space is capable of detonating. Hydrogen is released from storage batteries during

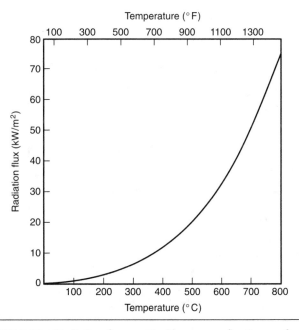

FIGURE 8.10 Radiation flux emitted by a nonreflecting surface versus temperature.

charging, and hydrogen is generated when acids attack metals. In addition, sodium or potassium reacts with water to form hydrogen.

Acetylene (C_2H_2)

Acetylene is an extremely reactive, flammable gas that cannot be stored in a compressed state alone without a possibility of dissociation into carbon and hydrogen, with release of energy. It is stored in cylinders that contain a very porous monolithic mass made of cement, asbestos, diatomaceous earth, and charcoal. The anhydrous filler mass, containing about 80 percent void space, is soaked with acetone. Acetylene gas is pumped into the cylinder and its filler from heavy, small-diameter piping (to withstand decomposition pressure, should it occur) until a maximum cylinder pressure of 250 psig (17 bar) is reached. The acetone dissolves 25 times its own volume of acetylene for each 14.7 psig (1 bar) pressure. An ordinary welding-type acetylene cylinder contains 5.5 gal (18.5 L) of acetone and about 20 lb (9 kg) of acetylene.

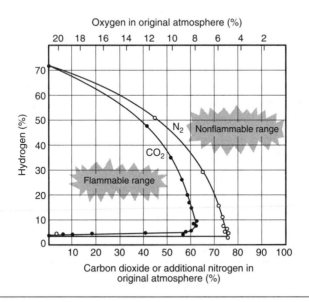

FIGURE 8.11 Influence of added inert gas on limits of flammability of hydrogen in air (downward propagation of flame).

Acetylene gas is exceedingly flammable, with a flammable range of 2.5 percent to 81 percent. Under certain conditions it will dissociate at gas concentrations from 81 percent to 100 percent, releasing heat energy in the process.

Because of the reactivity and unconventional storage method of acetylene gas, all acetylene tanks are provided with fusible plugs that open at about 212°F (100°C); should this occur near an ignition source, a flaming torch of burning gas will issue some distance — 10 to 12 ft (3 to 3.6 m) — from the opened vent. This flame need not be extinguished unless it endangers combustibles nearby. After a short time the torch will die down and the cylinder will cool sufficiently to be moved to a safe place where it can continue to vent. Some possibility exists that the flame will propagate back into the cylinder or tank, at which time the tank will heat up and must be cooled with water sprays. Danger of explosion of the tank exists only if it is heated to a glowing red color.

Acetylene–air flames propagate very rapidly, and an acetylene–air mixture in a suitably confined space is capable of detonating. Furthermore, acetylene in contact with copper forms copper acetylide, an extremely unstable solid that can explode.

Methane (CH$_4$)

Methane is an odorless gas, somewhat lighter than air. It is the main constituent of natural gas, to which odorants (sulfur compounds) generally are added as an aid to detection. Methane also is found in coal mines and is a major cause of coal mine explosions. Methane's flammability characteristics are similar to those of other saturated hydrocarbon gases (ethane, propane, n-butane, isobutane), all of which are odorless. (*See Figures 8.3 and 8.4.*)

Ethylene (C$_2$H$_4$)

Ethylene, also called ethene, is used widely as an industrial gas. It has a faint sweet odor. Its mixtures with air have wider flammability limits, are easier to ignite, and propagate flame more rapidly than saturated hydrocarbon gases. Ethylene flames are more luminous (sooty). Its mixtures with air can detonate, but not as readily as acetylene or hydrogen.

Ammonia (NH$_3$)

Ammonia is used widely as a commercial refrigerant and a fertilizer. It is not commonly known that some of its mixtures with air are flammable. An ammonia leak is easily detectable because of its sharp pungent odor. Ammonia–air mixtures are more difficult to ignite and burn more slowly than saturated hydrocarbon–air mixtures. (*See Figure 8.12.*)

FIGURE 8.12 Anhydrous ammonia vapor escaping from a pressurized container.

Problems

1. Are the flames on the burner of a gas stove premixed flames or diffusion flames? Why?

2. Why is turbulence important in combustion?

3. Is it easier to ignite a hexane vapor–air mixture or a hydrogen–air mixture? Does it depend on whether a spark or a hot surface is the ignition source?

4. How would one use the word *inflammable?*

5. Why are combustible gas–air mixtures flammable in certain proportions and not flammable in other proportions?

6. What is the burning velocity of a gas mixture? Under what circumstances can a flame move through a mixture faster than its burning velocity?

7. Why are fire flames usually yellow or orange? What is the practical significance of these colors?

8. What is a branching chain reaction?

9. Define smoke-point height. What is a possible use of this quantity?

10. What fraction of the energy of a yellow flame is convected upward, and what fraction is radiated in all directions?

11. Give three reasons why the thermal radiation from a flame is of importance in fire safety.

12. Give five reasons why an uncontrolled release of hydrogen creates a greater fire hazard than a comparable release of natural gas (chiefly methane).

13. What is a detonation? How does it differ from a deflagration?

References

1. Kuchta, J. M., "Investigation of Fire and Explosion Accidents in the Chemical, Mining, and Fuel-Related Industries — A Manual," Bulletin 680, U.S. Bureau of Mines, Washington, D.C., 1985.

2. Jost, W., *Explosion and Combustion Processes in Gases,* McGraw-Hill, New York, 1946.

3. Mullins, B. P., *Spontaneous Ignition of Liquid Fuels,* Butterworths, London, 1955.

4. *Fire Protection Handbook,* 16th ed., National Fire Protection Association, Quincy, MA, 1986, Section 8, p. 34.

5. Zabetakis, M. G., "Flammability Characteristics of Combustible Gases and Vapors," Bulletin 627, U.S. Bureau of Mines, Washington, D.C., 1965.

6. Glassman, I., *Combustion,* 3rd ed., Academic Press, New York, 1996.

7. Drysdale, D., *An Introduction to Fire Dynamics,* 2nd ed., J. Wiley, New York, 1998.

8. Strehlow, R. A., *Combustion Fundamentals,* McGraw-Hill, New York, 1984.

9. *Twenty-first Symposium (International) on Combustion,* The Combustion Institute, Pittsburgh, PA, 1988.

10. *Twenty-second Symposium (International) on Combustion,* The Combustion Institute, Pittsburgh, PA, 1989.

11. Schug, K. P., Y. Manheimer-Timnat, P. Yaccarino, and I. Glassman, "Sooting Behavior or Gaseous Hydrocarbon Diffusion Flames and the Influence of Additives," *Combustion Science and Technology,* Vol. 22, 1980, p. 235.

12. Hunt, R. A., "Relation of Smoke Point to Molecular Structure," *Industrial & Engineering Chemistry,* Vol. 45, 1953, pp. 602–606.

13. Parker, J. F., and V. R. West, *Bioastronautics Data Book,* 2nd ed., National Aeronautics and Space Administration, Washington, D.C., 1973, p. 114.

14. Fishburne, E. S., and H. S. Pergament, "The Dynamics and Radiant Intensity of Large Hydrogen Flames," *Seventeenth Symposium (International) on Combustion*, The Combustion Institute, Pittsburgh, PA, 1979, pp. 1063–1073.

15. Markstein, G. H., "Relationship Between Smoke Point and Radiant Emission from Buoyant Turbulent and Laminar Diffusion Flames," *Twentieth Symposium (International) on Combustion*, The Combustion Institute, Pittsburgh, PA, 1985, pp. 1055–1061.

16. Tewarson, A., "Generation of Heat and Chemical Compounds in Fires," *The SFPE Handbook of Fire Protection Engineering*, 2nd ed., National Fire Protection Association, Quincy, MA, 1995, Section 3, Chapter 4, pp. 3-83 – 3-124.

CHAPTER 9

Fire Characteristics: Liquid Combustibles

CATEGORIES OF LIQUID FIRES

The two most important questions with regard to liquid fires are as follows:

1. Which liquids are flammable?
2. How should flammable liquid fires be handled?

Five categories of liquid fires must be considered before these questions can be answered:

1. A pool of liquid, such as an open tank or the result of a spill *(See Figure 9.1.)*
2. A flowing liquid, such as an overflowing or rapidly leaking tank
3. A spray (requires high pressure and a small orifice), for example, a leaking hydraulic fluid line
4. A thin film: a foam or froth, or a thin liquid layer drawn up by capillarity over the surface of a fabric or paper, that is, wicking action
5. A confined liquid (in a pressure vessel, heated by an external fire), for example, an LPG tank car

For the first two categories, stagnant and flowing bulk liquids exposed to air, the question of flammability is settled easily. If the temperature of the liquid is above its *fire point* (which is explained later in this chapter), it can be

FIGURE 9.1 Flammable liquid spill.

ignited easily. With the proper equipment, fires on stagnant liquids are relatively easy to extinguish, by applying aqueous foams. Flowing liquid fires are
very difficult and sometimes impossible to extinguish, as long as the liquid
continues to flow.

Fires involving liquids in the form of sprays or thin films are different,
because the fire point temperature is no longer a relevant measure of flammability. A pool of domestic (No. 2) fuel oil at 20°C (68°F) cannot be ignited with
a match (unless a wick is present), but the same oil, in the form of a spray,
mist, or thin film, can be ignited easily.

A flammable liquid in a pressurized tank can pose a special kind of fire
threat. Consider a railroad car containing *liquefied petroleum gas* (LPG), which
is mainly a mixture of propane, *n*-butane, and isobutane. The pressure in the
tank will depend on the temperature and corresponding vapor pressure of the
contents and could be approximately 10 atm under normal conditions. A
pressure-relief valve is provided, which opens if the liquid overheats and generates an excessive pressure.

Now, assume that a fire starts outside the tank, either associated with LPG
being vented from the tank or because of an external combustible. The heat of
the fire acting through the tank wall will boil the LPG and generate vapor,
which will escape through the relief valve. As the wall of the steel tank gets
hot, its tensile strength diminishes. At about 500°C (932°F) the yield point of
steel is approximately one-half its normal value. At this point the tank might
rupture. The liquid contents, once boiling at perhaps 10 atm, are now suddenly
at 1 atm. An enormous eruption of vapor occurs, producing a *boiling liquid
expanded vapor explosion* [BLEVE (pronounced "blevey")]. *(See Figure 9.2.)* A
phenomenon similar to the BLEVE can occur with an ordinary cylinder of liquefied propane, such as one used for cooking or space heating in mobile
homes.

FIGURE 9.2   The fireball formed in a BLEVE involving LPG railroad tank cars. The elevated water tank at lower right is a point of reference in visualizing the tremendous dimensions of the fireball. (Courtesy of Anderson Watseka.)

Liquid fires also can be categorized according to whether the flammable liquid is highly water-soluble (e.g., acetone or alcohols) and therefore can be diluted with water, or whether the liquid is not water-soluble (e.g, gasoline) but floats on water. These characteristics have obvious implications in fire fighting.

FLASH POINT AND FIRE POINT

In general, flammable liquids do not burn. Their vapors burn. If an ignition source, such as a match flame, is placed just above a pool of flammable liquid, no ignition will occur unless the vapor pressure of the liquid is high enough to form a flammable vapor concentration higher than the lower flammable limit of that vapor in air.

Table 3.1 (p. 28) shows how the vapor pressures of some common liquids increase sharply with increasing temperature. Consider the data for methanol in Table 3.1. At 0°C (32°F), the vapor pressure is 29.7 mm of mercury (Hg). The total pressure is 760 mm Hg (1 atm), so the volume fraction, or mole fraction, of methanol vapor in the air just above the liquid surface is 29.7/760 = 0.039, or 3.9 percent by volume. Refer now to Table 8.3 (p. 93). The lower limit of flammability of methanol vapor in air is 6.7 percent by volume. Therefore, methanol should not be flammable at 0°C.

However, if a match flame was held next to a small enough quantity of methanol liquid for a long enough time, the liquid would not remain at 0°C, but would rise to a higher temperature at which the vapor pressure would be high enough to form a flammable mixture. At that time, the liquid would ignite.

Imagine that an open cup of cold methanol is being heated gradually from below. The methanol can be tested for flammability by passing a small flame (from a match or tiny burner) across the liquid surface every 10 seconds. When the liquid reaches 11°C (52°F), a flame will move rapidly across the surface, consuming the methanol vapor above the surface. After a fraction of a second, assuming the ignition source has been removed, there will be no further combustion. The reason is that the combustible vapor has been consumed, and by the time additional vaporization can occur to restore the original vapor concentration, there is no longer an ignition source. The minimum temperature at which this behavior occurs (11°C for methanol) is called the *flash point* of the liquid.

Now, if a liquid is heated in a cup above its flash point temperature, and the ignition flame was applied from time to time, combustion will continue indefinitely after removal of the ignition source at a liquid temperature a little higher than the flash point temperature (usually 5°C–15°C higher). This temperature is called the *fire point* of the liquid. It is the temperature at which the vapor pressure is high enough to maintain a supply of vapor as fast as it is consumed by the flame.

The exact values of flash point and fire point temperatures measured in such tests depend on the following properties:

1. The size of the ignition source
2. How long it is held over the liquid each time
3. The rate of heating the liquid
4. The degree of air movement over the liquid

Nevertheless, the measured values, especially the flash point, are widely used as guides to safe handling of liquids. The flash point, being lower than the fire point, is a more conservative value to use. Refer to the *Fire Protection Handbook* [1] for more information.

Liquids can be divided into classes (which are divided further into subclasses), based on their flash points [2]:

1. Class I — Liquids with flash points below 38°C (100°F)
2. Class II —Liquids with flash points at or above 38°C (100°F), but below 60°C (140°F)
3. Class III —Liquids with flash points at or above 60°C (140°F)

Assume that a liquid spill occurs on a summer day when the ground has been heated by the sun to 35°C (95°F). Clearly, a spill of a Class I liquid is extremely hazardous with regard to fire; a spill of a Class II liquid is dangerous from a fire viewpoint only if a heat source exists that is capable of raising the liquid temperature moderately; and a spill of a Class III liquid is safe from ignition unless a heat source exists that can raise its temperature substantially.

Table 9.1 shows flash points for some common liquids. Notice the wide range, from –43°C to +243°C. These values are meaningful only for a bulk liquid. If a liquid with a high flash point is in the form of a spray, a froth, or a foam, with air present, and comes into contact with even a very small ignition flame, the tiny amount of liquid in contact will be heated immediately to above its flash point and will start burning. The combustion energy released will vaporize the surrounding spray or foam, and the fire will propagate (spread).

TABLE 9.1   Flash points of some common liquids

	Flash point	
	°C	°F
Class I Liquids		
Gasoline	–43	–45
n-Hexane	–26	–15
JP-4 (jet aviation fuel)	–18	0
Acetone	–16	3
Toluene	9	48
Methanol	11	52
Ethanol	12	54
Turpentine	35	95
Class II Liquids		
No. 2 fuel oil (domestic)	>38	>100
Diesel fuel	40–55	104–131
Jet A (jet aviation fuel)	47	117
Kerosene	52	126
No. 5 fuel oil	>54	>130
Class III Liquids		
JP-5 (jet aviation fuel)	66	151
SAE No.10 lube oil	171	340
Tricresyl phosphate	243	469

Source: Henry [1] and Kuchta [3].

Suppose a high flash point bulk liquid has a wick projecting from it. The wick can be any nonmelting porous material that the liquid is capable of wetting. The wick can consist of a bit of cloth, paper, cardboard, and so on, that is in contact with the pool of liquid. (A discarded cigarette might simultaneously serve as a wick and an ignition source.) The liquid is drawn up the wick by surface tension (capillarity), and the wick becomes covered with a thin film of the liquid. (For example, immerse one corner of a handkerchief in a glass of water and observe what happens.)

If an ignition source is applied to the wick (such as a match to a candle wick), the thin film of liquid is heated rapidly to above its fire point and it ignites. As it burns, additional liquid is drawn up the wick and feeds the fire. Of course, such a fire is trivial in size. However, if this small fire is in contact with a large pool of high-flash-point liquid, its heat could eventually warm the liquid immediately adjacent to it so that the fire would spread from the wick to a portion of the liquid, and ultimately grow to a large fire.

BURNING RATES OF LIQUID POOLS

A pool of burning liquid will burn at a more or less steady rate from shortly after ignition until the liquid is consumed. To illustrate the magnitude of this rate, a pool of gasoline or mineral spirits 1 m (3.3 ft) in diameter and 2.5 cm (1 in.) deep will be consumed in about 4 minutes.

The rate of burning of a liquid is often expressed as a linear burning *rate* (in cm/min). This rate is readily converted to a mass burning rate (in g/cm^2-s) by multiplying the linear burning rate by the density of the liquid (in g/cm^3) and dividing the product by 60 (seconds per minute). Then, if the heat of combustion of the liquid (in J/g) is ascertained from a handbook, the rate of heat release per unit area can be calculated, assuming complete combustion. (See Table 8.5 on p. 104 for information on completeness of combustion.)

The linear burning rate of a pool of liquid will depend not only on the nature of the liquid but also on the diameter of the pool. Figure 9.3 shows the linear burning rate of *n*-hexane versus pool diameter. A progressive increase in linear burning rate is seen, although the rate of increase of burning rate is progressively less as the diameter is increased.

The reason for this increase becomes apparent if the factors that control the burning rate of a liquid pool are understood. Each gram of *n*-hexane that burns releases 44,860 J of heat energy. The rate of burning of the hexane is controlled by its rate of vaporization. In order to vaporize *n*-hexane, the latent heat of vaporization of 371 J/g must be supplied. Therefore, a little less than 1 percent of the combustion energy must return to the *n*-hexane surface, through the rising vapors, in order to maintain the fuel supply to the flame.

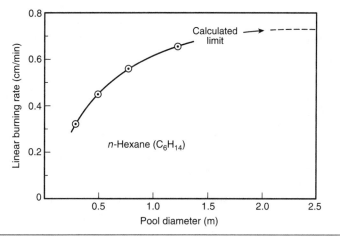

FIGURE 9.3 Effect of pool diameter on linear burning rate of *n*-hexane. Source: Kuchta. [3]

(The bulk of the energy is convected upward with the combustion products, or radiated sideways or upward.)

Detailed studies have shown that the energy transfer from a flame to a liquid surface is primarily by infrared radiation from the flame, at least for hydrocarbon pool fires with diameter larger than about 20 cm (8 in.). The flame, like any other semitransparent emitter of radiation, theoretically emits radiation as shown in Figure 9.4. Figure 9.4 shows that the intensity of radiation emerging from a flame of a given temperature increases linearly with the thickness of the flame for *optically thin* flames and levels off to a constant value for *optically thick* flames. The thickness required to reach the optically thick condition depends on how sooty the flame is. (See Radiation from Flames in Chapter 8, p. 102.) The set of curves in Figure 9.4 for a particular temperature is shifted to a higher level (proportional to the fourth power of absolute temperature) if the flame temperature is higher. Consequently, a pool of *n*-hexane burning in pure oxygen would burn much more rapidly than in air, because the flame temperature would be much higher.

Refer again to Figure 9.3 to see that the linear burning rate (or mass burning rate per unit area) is twice as high for a 1.2-m diameter *n*-hexane pool as for a 0.3-m diameter pool. The 0.3-m flame is nearly optically thin, the 1.2-m flame is approaching the optically thick condition. If the curve in Figure 9.3 is matched to a curve such as one of those shown in Figure 9.4, the linear burning rate for *n*-hexane is found to level off at 0.73 cm/min for pools larger than about 3 m.

Linear burning rates have been measured for other liquids and have been extrapolated to limiting values for large pool sizes. Refer to Kuchta [3] for a

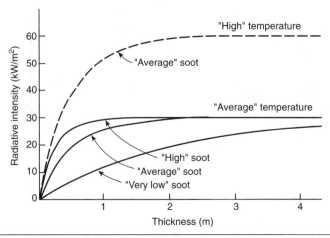

FIGURE 9.4 Calculated radiative intensity coming from a hot semi-transparent sooty gas of various thicknesses, for three soot levels and two temperatures. The "high" temperature is 19 percent higher than the "average" temperature, and the "high" soot level is 8 times the "very low" soot level.

tabulation. For large pools where the flames are optically thick, one might expect that the linear burning rate would be inversely proportional to the latent heat of vaporization of the liquid. Indeed, Figure 9.5 shows a correlation of linear burning rate versus the reciprocal of latent heat of vaporization per unit volume (the product of latent heat of vaporization and density) for five liquids.

The slope of the correlation line, divided by 60, turns out to be about 3 J/s/cm^2, or about 3 W/cm^2, or about 30 kW/m^2. This result implies that the radiation from optically thick flames of any of the five combustibles shown in Figure 9.5 imposes a heat flux of about 30 kW/m^2 on the liquid surface, regardless of the chemical nature of the combustible. The vast differences in sootiness of the various flames does not matter, since the flames are all optimally thick.

FLAME-SPREAD RATES OVER LIQUID SURFACES

The previous section dealt with burning rates of burning liquid pools, where the flame covers the entire surface. Now consider the rate of spread of the flame over the surface, just after one local region of the surface has been ignited.

Figure 9.6 shows data for *flame-spread rate* over the surface of a liquid, *n*-butanol (C$_4$H$_9$OH), which has a flash point of 43°C (110°F), measured by

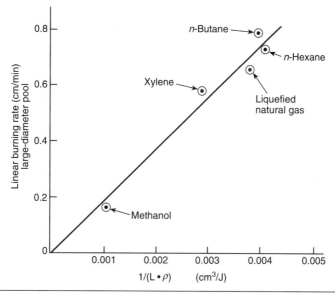

FIGURE 9.5 Burning rates of large liquid pools versus inverse of product of latent heat of vaporization (L) and density (ρ).

the open-cup method. Above this temperature, the flame-spread rate is 2 m/s (6.5 ft/s) and is independent of the liquid temperature. Below 43°C, the flame-spread rate is highly dependent on the liquid temperature. At 20°C the flame-spread rate is only $^1/_{100}$ of the value at 50°C.

This behavior is typical of other fluids. Flame will spread, but very slowly, over a liquid well below its flash-point temperature. In order to start the burning, the liquid must be heated locally to above its flash point. Then, the flame heats the adjacent cold liquid and induces convection currents in it, so that the flame spreads.

For fluids well above their flash-point temperatures, combustible vapor concentration in excess of the lower flammability limit will exist over the surface before the arrival of the flame. If the liquid is warm enough, the vapor concentration just over the surface could be above the upper flammability limit. However, if normal air circulation exists over the liquid pool, the combustible concentration will decrease with increasing height above the surface, and somewhere there will be a zone containing a near-stoichiometric mixture. The flame will propagate across the surface through this zone, at a speed several times the burning velocity. (See Burning Velocity in Chapter 8, p. 93.)

For a stoichiometric n-butanol–air mixture, the burning velocity is about 0.6 m/s. The flame propagates about three times this fast because, instead of

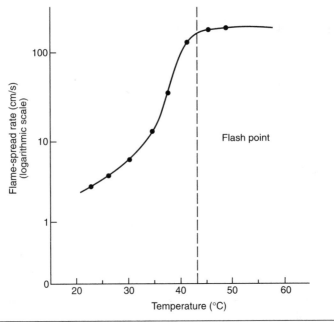

FIGURE 9.6 Rate of flame spread over the surface of *n*-butanol.
Source: Burgoyne and Roberts. [4]

being vertical, it is inclined at an angle in the flammable zone. If a combustible fluid is spilled to cover the floor of a room, a propagation speed of 2 m/s would mean that a flame could spread completely across a normal-size room in no more than 2 or 3 seconds. A breeze would, of course, have a profound effect on the flame-spread rate and would carry the flame downwind.

BOILOVER

The density of a liquid generally decreases slightly with increasing temperature. Consider a fire burning on the surface of a deep pool of liquid in an open tank. The liquid just below the surface will be warmer and therefore lighter than the liquid below, so there will be very little mixing between the upper and the lower portions of the liquid. The burning will proceed at a fairly steady rate, unless the fire is extinguished.

Imagine a fire in an open tank of crude oil, with a layer of water on the bottom of the tank (which is often the case). Crude oil consists of a mixture of highly volatile and slightly volatile components. A heated sample would start to burn below 100°C, but the temperature of the sample would rise to over 300°C by the end of boiling. This temperature range is referred to as the distillation range.

As the fire burns in the open tank, the more volatile components will be driven out of the topmost layer of liquid, and the temperature of this layer will rise to about 300°C. Heat will be conducted downward through the liquid to the next layer, causing gasification of volatile components in that layer, with bubble formation. The motion of these bubbles will greatly accelerate the mixing of the hotter upper fluid and the cooler lower fluid. Because of this bubble-induced mixing, a "hot zone" will move down through the liquid and eventually reach the water at the bottom. If conditions are such that the hot oil contacting the water is well above 100°C, the water will start to boil vigorously. An oil-steam froth will form at the surface, and this burning froth could suddenly overflow the tank, with disastrous consequences to any nearby fire fighters or spectators. Such incidents have occurred in refinery fires. The phenomenon is called *boilover*.

Boilover has not been understood fully until recently. The foregoing mechanistic explanation was worked out by Hasegawa [5], based on careful experiments. Boilover will occur only in an open tank containing a blend of flammable liquids with a wide distillation range; water also must be present at the bottom of the tank.

Problems

1. Describe how one can predict the fire point of a liquid if its vapor pressure versus temperature and all the combustion properties of its vapor are known.

2. Under what conditions can a liquid burn if its temperature is below its flash point?

3. Define the linear burning rate of a pool of liquid. How does this rate depend on the size of the pool?

4. How does the rate of flame spread over the surface of a liquid depend on the flash point?

5. Under what conditions could boilover occur?

6. Under what conditions could a BLEVE occur?

References

1. Henry, M. F., "Flammable and Combustible Liquids," *Fire Protection Handbook*, 16th ed., National Fire Protection Association, Quincy, MA, 1986, Section 5, pp. 28–38.

2. Slye, O. M. Jr., "Flammable and Combustible Liquids," *Fire Protection Handbook,* 18th ed., National Fire Protection Association, Quincy, MA, 1997, pp. 4–60.

3. Kuchta, J. M., "Investigation of Fire and Explosion Accidents in the Chemical, Mining, and Fuel-Related Industries — A Manual," Bulletin 680, U.S. Bureau of Mines, Washington, D.C., 1985.

4. Burgoyne, J. H., and A. F. Roberts, "The Spread of Flame Across a Liquid Surface," *Proceedings of the Royal Society, London,* 1968, Vol. A308, pp. 39–79.

5. Hasegawa, K., "Experimental Study on the Mechanism of Hot Zone Formation in Open-Tank Fires," *Fire Safety Science: Proceedings of the Second International Symposium,* Hemisphere Publishing Company, New York, 1989, pp. 221–230.

CHAPTER 10

Fire Characteristics: Solid Combustibles

GASIFICATION, IGNITION, CHARRING, AND MELTING

Organic solids (carbon compounds) must gasify before they can burn. This general principle, however, is not true of carbon itself, which burns by a surface reaction with oxygen (although the carbon monoxide that forms in this reaction might burn with oxygen in the gas phase). With the exception of carbon, the burning process of ordinary solids is preceded by a gasification process, generally induced by heat and requiring the breaking of chemical bonds. The subsequent fuel–oxygen combustion reaction occurs in the gas phase. The term *pyrolysis* is used to describe heat-induced chemical decomposition resulting in gasification.

Pyrolysis is usually endothermic; that is, a given quantity of heat must be supplied from somewhere for the reaction to occur. This quantity of heat is the *heat of gasification* of a material, which is expressed in kJ/g. It is an important measure of the flammability of a solid, once ignited.

The minimum condition for igniting a solid is the heating of its surface to a temperature high enough that the pyrolysis gases are produced rapidly enough to exceed the lower flammability limit above the surface. (Any vapors produced are being diluted continuously by mixing with air.) This condition usually requires a gasification rate of a few grams per square meter per second.

Chemical kinetic principles show that the pyrolysis reaction rate at the solid surface increases rapidly with increasing temperature. *(See Chapter 3.)*

For most organic solids, a temperature between 270°C and 400°C (518°F and 752°F) is needed.

Even when the necessary vapor concentration has been achieved, ignition will not occur unless an ignition source is present, such as a flame or a spark. In the absence of an ignition source, ignition is still possible if the surface can be heated to such a high temperature that *spontaneous ignition* occurs. For example, when wood is heated with a radiant heater, *piloted ignition* (initiated by a small flame maintained near the surface) occurs when the wood surface reaches 300 to 400°C (572 to 752°F), while the same surface must be heated to about 600°C (1112°F) to induce spontaneous ignition.

Even if a surface is heated to the necessary temperature and a pilot flame is present, causing ignition, the flame might self-extinguish shortly after the external heat source is removed. Whether or not the flame is extinguished depends on two key factors, both related to heat loss from the surface:

1. The surface can lose heat by conduction into the interior of the solid (unless it is thin and/or has been heated throughout).

2. The surface can lose heat by radiating it away to cooler surroundings.

Therefore, if a thick solid is heated very gradually, when the surface reaches the ignition temperature, the interior is already quite hot and will not drain off heat very fast from the surface. Thus, the pre-ignition heating rate is important in achieving sustained ignition of thick solids. If the solid is pre-heated adequately all the way through, there is no question that it will continue to burn after ignition.

Now, consider radiative heat loss. If a lighted match is placed between two massive pieces of wood facing each other, for example, 1.3 cm (0.5 in.) apart, then sustained ignition is achieved easily because the radiative heat lost from each surface is captured by the other surface and there is no net loss. *(See Figure 10.1.)* Achieving sustained ignition of a single thick piece of wood with a match usually is not possible because of radiative heat loss to the surroundings.

The minimum intensity of radiative flux that must impinge on a solid to make it ignitible by a pilot flame has been measured for many materials. These values range from 10 to 40 kW/m^2, depending on the nature of the material, including its:

1. Chemical constituents

2. Reflectivity

3. Size

4. Orientation

For fluxes in excess of the minimum value, the time to ignition decreases as the flux increases. *(See Figure 10.2.)*

FIGURE 10.1 Sustaining of ignition. The radiative heat from one surface is captured by another surface.

After a solid has been ignited and the flame has spread across its surface, two distinct categories of burning behavior are apparent. One class of material, including wood and certain plastics, burns with the formation of a growing char layer. The other class of material, consisting of many of the more common plastics (e.g., polyethylene, polystyrene, and Plexiglas®), burns with either no char or a small amount of surface char that never builds up to a thick layer. The importance of char formation is seen in Figure 10.3, which illustrates heat-release rate versus time for a char-forming (particleboard) and a nonchar-forming (Plexiglas) material.

The porous char, which forms as a thermal decomposition product of many organic solids, is not quite pure carbon; it contains 1 hydrogen atom for every 5 or 6 carbon atoms, which are arranged in adjacent six-membered rings (polycyclic structure). The char is not a good conductor of heat, and as it grows thicker it progressively slows down the rate of endothermic pyrolysis of the underlying material. Therefore, the rate of burning is reduced. The char ultimately develops cracks and fissures. *(See Figure 10.4.)*

The rate of char formation of wood has been reported to be proportional to the radiant heat flux impinging on the surface. [1] For a "typical" radiant heat flux of 30 kW/m², which might exist just under a flame, the average char-ring rate would be about 0.6 mm/min (0.025 in./min).

Refer again to Figure 10.3, to see that a wood (particleboard) sample, exposed to a radiative flux of 25 kW/m², burns rapidly for a short time, after which the burning rate progressively slows down to a much lower rate as the

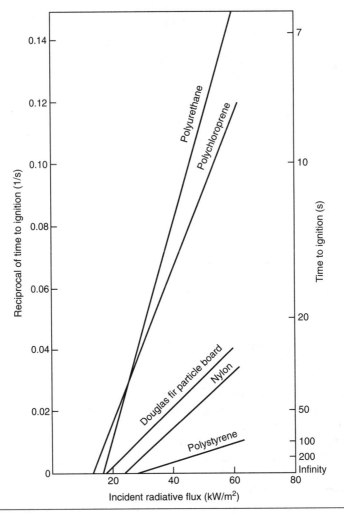

FIGURE 10.2 Effect of radiative flux intensity on time to achieve piloted ignition. Data (unpublished) from Factory Mutual Research Corporation, except data on Douglas fir, which is from U.S. National Institute of Standards and Technology.

char layer builds up. The final increase in the curve is attributed to heating up of the insulated back side of the sample by the time the combustion zone arrives there.

In contrast, the noncharring Plexiglas sample in Figure 10.3 burns at a high rate throughout the burning period, except at the beginning and the end.

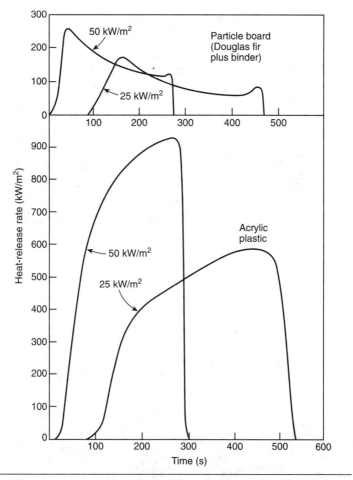

FIGURE 10.3 Heat-release rates vs. time for wood and Plexiglas samples under imposed radiative heat fluxes of 25 and 50 kW/m².

This high burning rate makes noncharring combustibles more dangerous than charring combustibles.

As one would expect, if the incident radiative flux is increased to a higher value, both the charring and the noncharring materials burn faster. Some combustibles, primarily noncharring ones, will melt while burning. In some cases (e.g., Plexiglas), the melt is very viscous, and little flowing occurs; in other cases, the melt is watery in consistency (e.g., polyethylene, polypropylene, and polystyrene) and, if the physical arrangement permits, burning drops of molten plastic will fall or flow downward and provide a means of spreading the fire as shown in Figure 10.5.

FIGURE 10.4 Charred ceiling beams.

FIGURE 10.5 Molten plastic from overhead light fixtures.

FLAMMABILITY, FLAME-SPREAD RATES, AND BURNING RATES

Flammability

Although most organic materials are flammable, there are various degrees of flammability, regardless of whether it is measured by ease of ignition, rate of flame spread, rate of heat release (burning rate), or lowest oxygen concentration at which burning can occur. Much effort has been expended on developing small-scale laboratory tests to evaluate materials on a scale of flammability, but with limited success.

The radiation flux from the existing fire is very important in determining how much a given material in a compartment can contribute to a fire that is already consuming other materials in that compartment. There are three radiation sources:

1. The flames themselves
2. The hot smoke
3. The hot ceiling and upper walls

Small-scale tests that do not simulate the radiation tend to give misleading answers in flammability evaluations.

Ignition and burning-rate tests for solid samples subjected to radiation from electric heating elements have been developed. [2, 3, 4] Figure 10.6 gives an idea of how much an external flux of 10 or 20 kW/m^2 can accelerate the burning rate of a slab of material. In the case of wood, a single thick slab will not continue to burn except with an external radiative flux. Figure 10.6 also shows the vastly different heat-release rates of materials.

Flame-Spread Rates

Generally, a fire will originate at a point on the surface of a flammable material, and will spread from that point. The initiation is caused most commonly by a cigarette, a match, or a faulty or improperly located electric device or heating apparatus. Once the flame has progressed a short distance from the source, its further rate of flame spread will become independent of the source but will be dependent on a number of other variables:

1. Direction of spread
2. Radiative preheating
3. Thickness of the solid

The rate of fire spread in a horizontal or downward direction over the surface of a thick solid is generally very slow ("creeping flame spread"), of the order of a few hundredths of a millimeter per second (0.1 in. per min), unless the surface has been preheated. For example, if Plexiglas is preheated

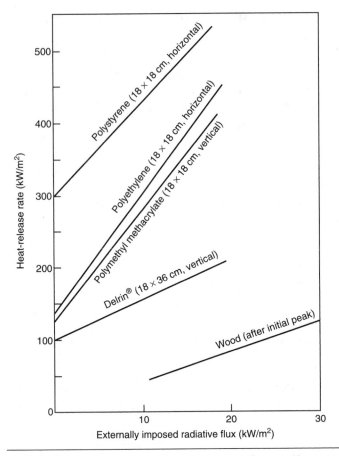

FIGURE 10.6 Heat-release rates of solids as influenced by externally imposed radiation. The magnitude of the heat-release rates per unit area will depend on the size and orientation of the sample. Data from Magee and Reitz [5] and Butler [1], plus combustion efficiencies from Tewarson [6].

for 1 minute with a radiative flux of 20 kW/m², the downward spread rate increases from 0.05 to 0.5 mm/s (a factor of 10). More prolonged or more intense preheating would produce further acceleration, by at least another factor of 10.

Flame spread in an upward direction is far more rapid than horizontal or downward spread, and, furthermore, the flame accelerates as it spreads upward. The reasons for this difference in behavior are as follows.

For downward or horizontal spread, the air is drawn into the base of the flame in a direction such that the flame must propagate upstream with respect to the air motion. This prevents the flame from getting very close to the unburned portion of the surface. Also, if the surface is heated to some degree by the flame radiation, it is simultaneously cooled by the approaching air. *(See Figure 10.7.)*

Conversely, for upward spread, the flame is in close contact with the not-yet-ignited portion of the combustible and can preheat it efficiently by both convection and radiation, so rapid upward spread can be expected. Furthermore, as the flame spreads upward, it becomes larger. Its greater length and thickness promote radiative heat transfer. Its greater length and higher gas velocity promote convective heat transfer. Thus, the upward spread rate increases progressively. On high walls, flame-spread speeds of several meters per second are possible.

Another powerful factor affecting flame spread, in addition to direction and radiative preheating, is the thickness of the solid. Downward spread experiments with thin cardboard and fabric samples have shown that the rate of spread is inversely proportional to the thickness. This behavior can be predicted from the concept that, in order to spread, the flame must heat the adjacent material to the ignition temperature. If the material is twice as thick, it will take twice as long to be heated, and the flame will advance one-half as fast. As a consequence of this principle, flame can spread very rapidly over a material composed of thin elements, such as a foam plastic, a pile of wood chips, or tissue paper.

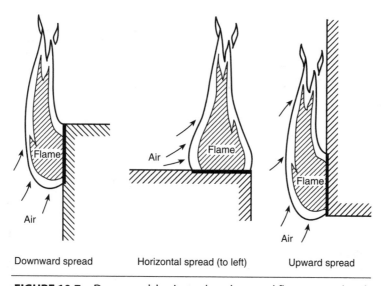

Downward spread   Horizontal spread (to left)   Upward spread

FIGURE 10.7 Downward, horizontal, and upward flame spread and direction of buoyancy-induced air flow at the base of the flame. The heavy line on the surface denotes the burning area.

Burning Rates

Provided ample air is available, once a flame has spread completely over the surface of a flammable item, combustion will continue at a rate dependent on the nature of the material and, for burning in a compartment, on the feedback of radiation from the hot smoke layer and from the ceiling. Nearby items can ignite either by radiation or by *firebrands* (airborne burning particles).

The rate of burning of a given material depends on its heat of gasification. Table 10.1 presents values for heats of gasification of a number of combustible solids. The values range widely, from 1.19 to 3.74 kJ/g. Furthermore, note that there is a variation in heats of gasification of samples of the same generic material from different sources. Presumably this variation is caused by chemical differences. If these heats of gasification (1.19 to 3.74 kJ/g) are compared with the heats of combustion of the same materials (which range from 15 to 44 kJ/g), it is apparent that only a small portion of the heat released by burning must return to the pyrolyzing solid in order to maintain a continuing supply of combustible vapor to the flame.

Burning rate can be expressed in grams per square meter per second or kilowatts per square meter. When expressed as kilowatts per square meter, the burning rate also is referred to as *heat-release rate*. The steady burning rate of a given material under given conditions depends on a balance of energy at the surface. The rate of energy absorption per unit surface area is the product of

TABLE 10.1 Heat of gasification for selected solids

Material	Heat of gasification (kJ/g)	Number of materials tested
Acrylonitrile-butadiene-styrene polymer	3.23	1
Corrugated paper	2.21	1
Douglas fir	1.82	1
Nylon 6/6	2.35	1
Phenolic plastic	1.64	1
Polyester with glass fibers	1.39–1.75	2
Polyethylene	1.75–2.32	2
Polyisocyanurate foam	1.52–3.74	2
Polymethyl methacrylate (Plexiglas)	1.63	1
Polyoxymethylene (Delrin®)	2.43	1
Polystyrene, granular	1.70	1
Polystyrene foam	1.31–1.94	5
Polyurethane foam	1.19–2.05	7
Polyvinyl chloride, rigid	2.47	1

Source: Tewarson, Section 1, p. 180. [6]

the mass rate of gasification and the heat of gasification. This product must equal the energy flux from the flame to the surface, minus the rate at which energy is lost from the hot surface by reradiation to cold surroundings, if any.

One way in which the burning of solids differs from burning of liquids is that burning liquids generally have much lower surface temperatures, with little reradiation. Another difference is that liquids gasify without chemical change, while solids usually decompose. Otherwise, the burning processes of liquids and solids are similar.

Table 10.2 shows a detailed energy balance for a 30.5 cm by 31.1 cm (1 ft by 1 ft) slab of Plexiglas burning in a horizontal configuration. Table 10.2 shows that 73 percent of the heat transfer from the flame to the surface is by flame radiation, and 27 percent is by convection from the flame gases just above the surface. Also, 39 percent of the total heat transfer to the surface is reradiated from the surface. Thus, radiation is important for a fire of this size.

If the test piece of Plexiglas had been larger, then, because of a thicker flame, the flame radiation flux per unit area would have been even greater, and the burning rate per unit area would have been larger. Figure 10.8 illustrates this effect by showing burning rates for horizontal Plexiglas slabs of various sizes.

Given the data in Figure 10.8, what would happen to the burning rate per unit area if a *very* large slab of Plexiglas, say 5 m² (16 sq ft) were burned? The

TABLE 10.2   Energy balance for a horizontal Plexiglas slab burning in air

Conditions	
Flame base	0.305 m by 0.311 m
Mass burning rate	0.996 g/s
Surface temperature	385°C
Heat of gasification	1.61 kJ/g
Heat of combustion	24.9 kJ/g
Combustion efficiency	85 percent
Heat-release rate	21.1 kW
Radiative fraction	0.34
Heat flux from flame to surface	
By radiation	1.91 kW (73 percent)
By convection	0.72 kW (27 percent)
Total	2.63 kW
Heat absorbed by gasification	1.61 kW (61 percent)
Surface reradiation loss	1.02 kW (39 percent)
Total	2.63 kW

Source: de Ris, p. 1012. [7]

temperature in a very large flame would be about the same as in a smaller one, and, for a heat source at a given temperature, there is an upper limit to the radiant intensity it can generate. Therefore, it can be predicted that the rising curve in Figure 10.8 would level off at some point. This limit for Plexiglas is not known exactly, but it has been estimated to be between 30 and 60 g/m²/s.

Even for a piece of Plexiglas within the sizes shown in Figure 10.8, the burning rate will depend on other factors if they are present. If the Plexiglas were vertical instead of horizontal, it would burn more slowly because the flame would be thinner in the direction normal to the surface and would radiate less heat to the surface. If the Plexiglas were burning in a hot compartment with radiative feedback, say from the ceiling, or from another flame, the burning rate would increase. If the air entering the flame were diluted with pre-cooled combustion products, the burning rate would decrease. Or, on the other hand, oxygen enrichment of the air would increase the burning rate.

Finally, if the burning object were made of some plastic other than Plexiglas, this would influence the burning rate. For instance, a 0.305 m by 0.305 m slab of Delrin (polyoxymethylene) would release heat at less than one-half the rate of acrylic plastic (polymethyl methacrylate), while the same size slab of polystyrene would release heat 48 percent faster than acrylic and 3.29 times as fast as Delrin. [7] These differences are caused by differences in flame radiation, in surface reradiation, and in heat of gasification.

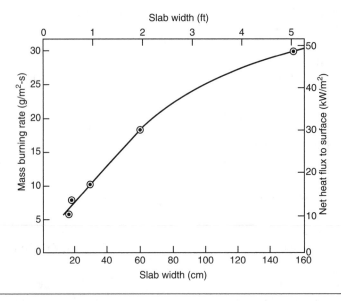

FIGURE 10.8 Mass burning rates of square, horizontal Plexiglas slabs of various sizes. Data from Factory Mutual Research Corporation (unpublished)

CELLULOSIC AND OTHER NATURAL MATERIALS

Wood, paper, and cotton, which are frequently involved in fires, are all cellulosic materials. *Cellulose*, which can be isolated in pure form from the natural material cotton, is a condensation polymer of glucose, $C_6H_{12}O_6$, a form of sugar. The chemical structures of glucose and cellulose are shown in Figure 10.9 (three-dimensional aspects are not shown).

There are generally more than 20,000 $C_6H_{10}O_5$ units in each molecule of cellulose. The name *condensation polymer* refers to the molecule of H_2O that must "condense out" each time another link is added to the polymer chain:

$$C_6H_{12}O_6 + C_6H_{12}O_6 \rightarrow C_{12}H_{22}O_{11} + H_2O$$

$$C_{12}H_{22}O_{11} + C_6H_{12}O_6 \rightarrow C_{18}H_{32}O_{16} + H_2O$$

If cellulose is boiled in acid, it decomposes, and glucose is a product.

Although cotton consists of more than 90 percent cellulose, wood contains only 40 to 50 percent cellulose. In the process of making paper from wood, much of the noncellulosic material is removed, and, therefore, the cellulose content of paper is fairly high; the exact value depends on the type of paper.

As is well known, there are many types of wood, with substantial variations in properties. For example, the densities of ponderosa pine (a softwood) and white oak (a hardwood) are 0.42 and 0.73 g/cm^3, respectively. Wood, which is somewhat porous in structure, can absorb large amounts of moisture, up to 25 percent in extreme conditions. After long contact with dry air at 20°C and 20 percent relative humidity, wood will have a moisture content of about 5 percent. The fire properties of wood, especially the ease of ignition, are influenced by the moisture content. The chemical composition of dry wood is as follows, varying somewhat with the type of wood:

40 to 50 percent cellulose

18 to 35 percent lignin

10 to 30 percent hemicellulose

5 to 20 percent "extractives" (oils, tars, gums, etc.)

0.2 to 1 percent minerals

Lignin is a *cross-linked polymer* that acts to bind the cellulose fibers together and add strength. (When wrapping paper is made from wood, not all the lignin is removed, so that the paper will be strong. Much more of the lignin is removed in making tissue paper.) When heated, lignin decomposes in a different manner than cellulose, which starts to decompose at about 250°C, and is mostly decomposed at 370°C, leaving a small amount of char behind. Lignin starts to decompose below 300°C, but even after prolonged heating at

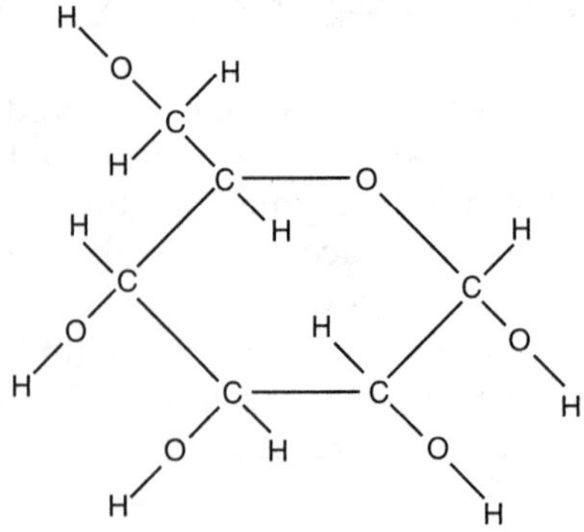

Glucose ($C_6H_{12}O_6$)

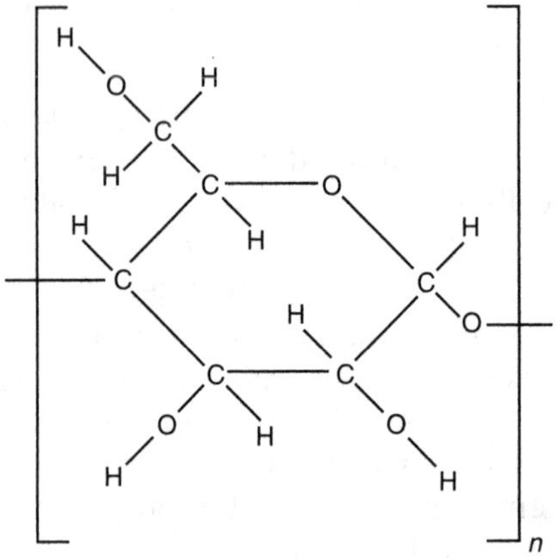

Cellulose ($C_6H_{10}O_5)n$
[$n > 20{,}000$]

FIGURE 10.9 The structures of glucose and cellulose.

high temperatures, in an inert atmosphere, about one-half of the original mass remains as char.

Hemicellulose is a polymer of glucose and other sugars, and has a much lower molecular weight than cellulose. One of its molecules can contain several hundred sugar units instead of tens of thousands.

Figure 10.10 shows how wood gasifies when heated slowly in an inert atmosphere. It is clear that the gasification occurs over a fairly wide range of temperatures, leaving a substantial char residue (15 to 25 percent of the original weight).

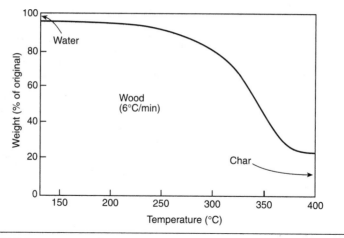

FIGURE 10.10 Gasification of wood during slow heating. Source: *Kirk-Othmer Encyclopedia of Chemical Technology,* p. 593. [8]

Chemical analyses of the volatile gases emerging from heated wood reveal the presence of hundreds of different chemical species. The major constituents are carbon monoxide, water vapor, and carbon dioxide. In addition, there are hydrogen, methanol, formaldehyde, acetaldehyde, acrolein, methane, ethylene, and so on. The relative proportions of various species depend on heating conditions as well as the type of wood. A number of the species are irritant or toxic.

The net heat of combustion of dry wood, approximately 18 kJ/g, is a composite of the heat of combustion of the volatiles produced during pyrolysis (approximately 14 kJ/g of volatiles) and the heat of combustion of the residual char (approximately 34 kJ/g of char). In an actual fire, the char could burn at a later time than the volatiles.

Other natural materials in general use include wool, leather, and silk. These are all animal products with high protein content. The proteins are polymer molecules with the monomer units (amino-acids) connected by peptide linkages:

$$\begin{array}{cc} O & H \\ \| & | \\ -C - N - \end{array}$$

Upon thermal decomposition, the gases evolved include ammonia (NH_3), amines (e.g., CH_3NH_2), and traces of hydrogen cyanide (HCN). Generally, these materials are appreciably less flammable than cellulosic materials, but they will burn under a high heat flux and will produce irritant and toxic gases in addition to the usual toxic species found in cellulosic fires.

SYNTHETIC POLYMERIC MATERIALS

A *polymer* is a very long molecule, consisting of a chain of thousands of atoms, created by reacting simple molecules called *monomers* with one another. Two monomers combine to form a *dimer*; three monomers combine to form a *trimer*, and so forth. These terms come from the following Greek combining forms.

mer = part

mono = one

di = two

tri = three

tetra = four

penta = five

hexa = six

oligo = several

poly = many

Monomers can combine to form either *addition polymers* (e.g., polyethylene) or *condensation polymers* (e.g., cellulose). An addition polymer contains the same atoms originally present in the monomer from which it was formed.

Addition Polymers

An important class of addition polymers is based on monomers that each have a double bond, for example, ethylene (C_2H_4). *(See Chapter 3, p. 32.)* Figure 10.11 shows a number of substituted ethylene monomers. Each of these monomers can form a commercially important vinyl polymer. The radical CH_2–$CH—$ is called vinyl, and, therefore, CH_2–CHX molecules are all vinyl monomers, although many of them have other chemical names.

Vinyl-type monomers can form *linear polymers* that consist of long chains of atoms, which, in spite of the term *linear*, are not arranged in a straight line. These long chains are twisted and coiled; however, there are no branches in the chains.

Branch points can be introduced and used to tie together adjacent strands to form a three-dimensional network if it is desirable to:

1. Reduce the flexibility

2. Increase the strength or hardness, or increase the melting point of a polymer.

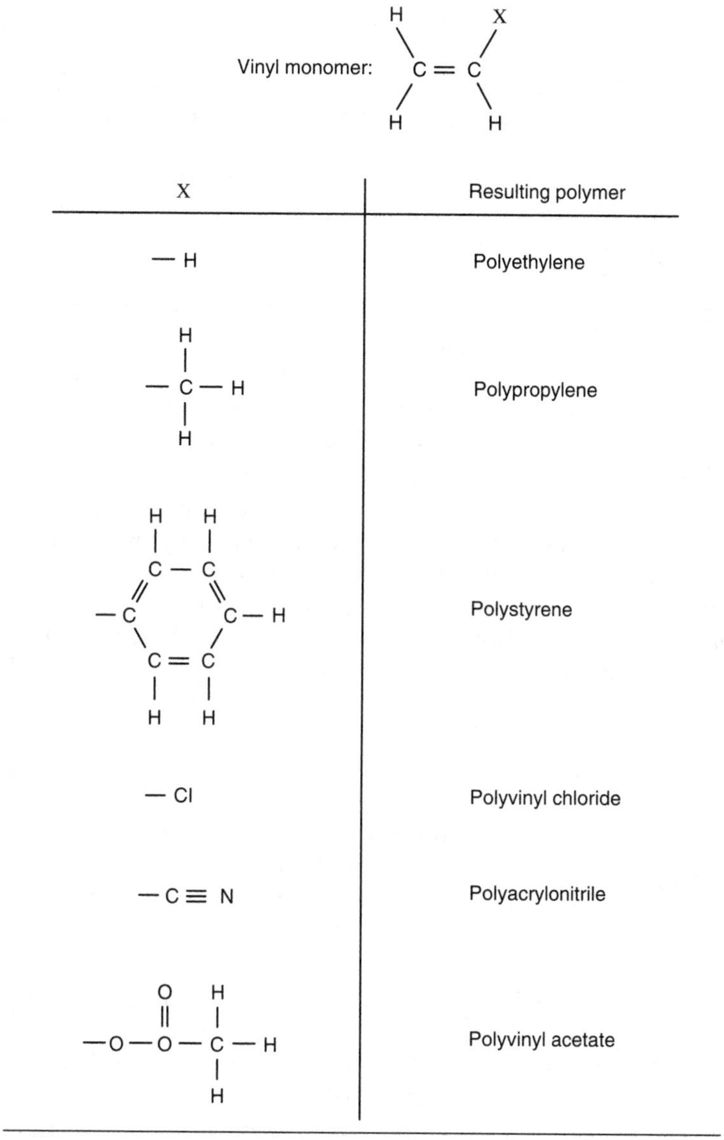

Vinyl monomer:

X	Resulting polymer
— H	Polyethylene
H | — C — H | H	Polypropylene
(ring structure)	Polystyrene
— Cl	Polyvinyl chloride
— C ≡ N	Polyacrylonitrile
O H || | —O—O— C — H | H	Polyvinyl acetate

FIGURE 10.11 Some vinyl monomers capable of forming vinyl polymers by linear addition.

This is called *cross-linking,* and, for vinyl polymers, is accomplished by adding a small percentage of a monomer containing two double bonds per molecule (a cross-linking agent) to the principal monomer. For example, if 1 percent of divinyl benzene,

$$CH_2=CH-(C_6H_4)-CH=CH_2$$

is added to styrene ($C_6H_5-CH=CH_2$) monomer, then linkages will form during polymerization that tie together nearby polystyrene chains at various random points. As another example, the "B" in the well-known ABS polymer is butadiene, $CH_2=CH-CH=CH_2$, which acts to cross-link the otherwise linear chains of the mixture of the two vinyl monomers acrylonitrile ("A") and styrene ("S"). Incidentally, this kind of polymer, formed from several kinds of monomers, is called a *copolymer*.

From the viewpoint of fire, the significance of cross-linking is that it reduces the flammability by raising the temperature required for gasification. This higher temperature promotes char formation, and the more char, the fewer combustible volatiles. Figures 10.12 and 10.13 show some monomers not involving the vinyl group, but which also lead to addition polymers.

When addition polymers are heated sufficiently, they all pyrolyze to form gases, but not all in the same way. Some of them, particularly polymethyl methacrylate, Delrin, and polytetrafluoroethylene, "unzip" from an end of each polymer chain and break off one monomer unit at a time, which immediately vaporizes. Other polymer molecules, such as polyethylene and polypropylene, break at random points to form fragments called *oligomers*, which vaporize if small enough. Thus, the pyrolysis vapors from polyethylene or polypropylene consist of a mixture of molecules having from 2 to 12 or more carbon atoms each.

Polyvinyl chloride (PVC) pyrolyzes differently from either of the above modes. At about 250°C the molecule HCl splits out from the chain and gasifies, leaving behind a char-forming material. Therefore, pure PVC has very low flammability. However, in pure form it is a rigid material, and substantial proportions of flammable plasticizers are often added to it to impart flexibility. In this way PVC can be used as electric cable insulation or vinyl furniture coverings.

Modacrylics are copolymers of acrylonitrile and either vinyl chloride or vinylidene chloride (CH_2Cl_2). The introduction of the chlorine atom improves the fire performance, relative to ordinary acrylic polymers. Of course, the HCl present in the fire gases is toxic and corrosive.

Fluorine substitution does not improve fire performance as much as chlorine substitution. Nevertheless, if all hydrogen atoms are replaced by fluorine, as in Teflon®, an almost nonflammable polymer is obtained. Even so, Teflon will burn in pure oxygen. Bromine substitution would be even more effective against fire than chlorine substitution.

Polyurethane is used widely in solid form and as foam, the foam being either flexible or rigid, with a wide variation of burning properties for the various polyurethanes. Isocyanurate cross-linked rigid foams, used for insulation, are less flammable than flexible foams.

Monomer	Resulting polymer
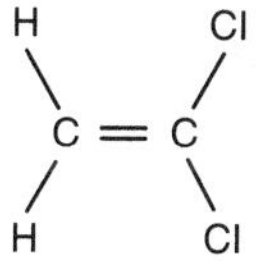	Polyvinylidene chloride (Saran)
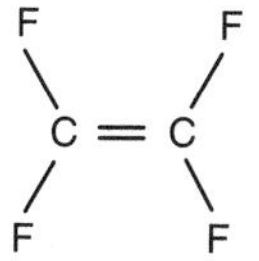	Polytetrafluoroethylene (Teflon)
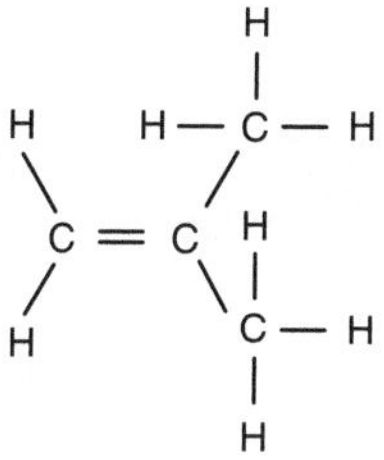	Polyisobutylene
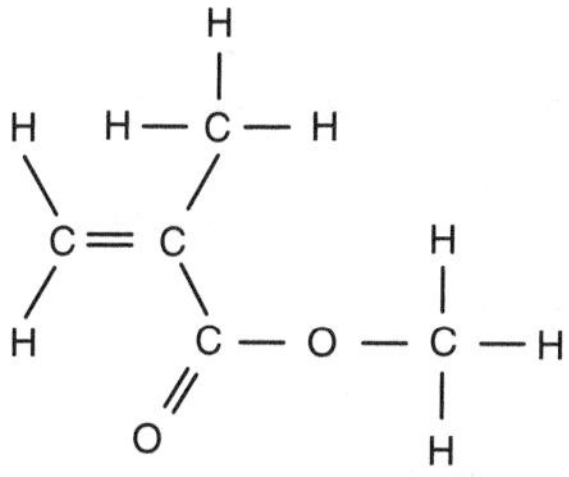	Polymethyl methacrylate (acrylic plastic)
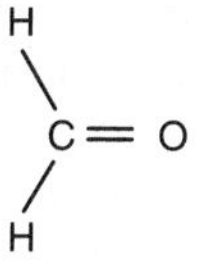	Polyoxymethylene (a polyacetal, Delrin) 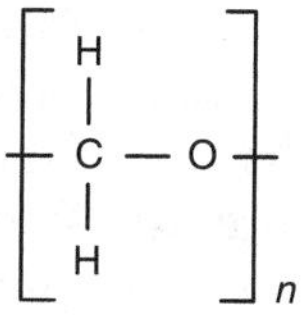
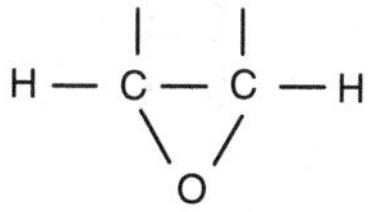	Polyethylene oxide (also a polyacetal) 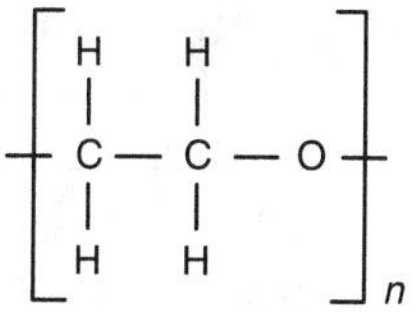

FIGURE 10.12 Some nonvinyl monomers capable of forming addition polymers.

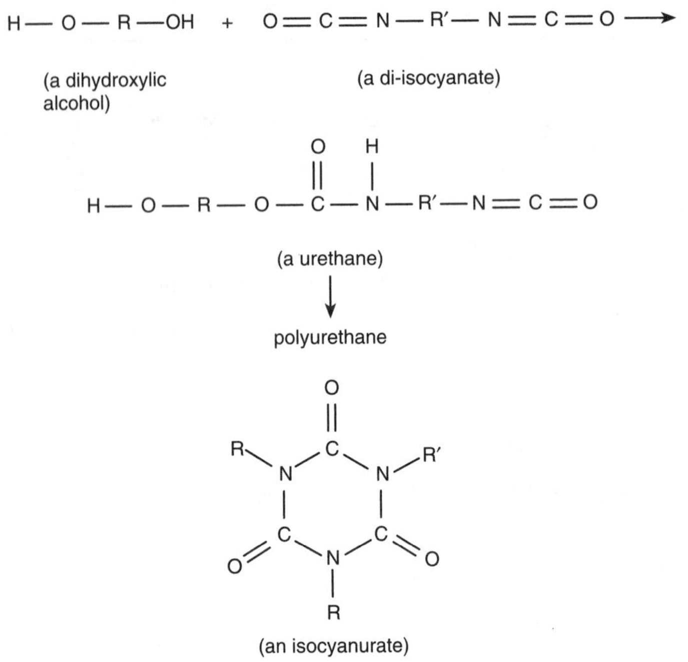

H—O—R—OH + O=C=N—R'—N=C=O ⟶

(a dihydroxylic alcohol)   (a di-isocyanate)

$$\underset{\text{(a urethane)}}{H-O-R-O-\overset{\displaystyle O}{\overset{\|}{C}}-\underset{}{\overset{H}{\underset{|}{N}}}-R'-N=C=O}$$

↓

polyurethane

(an isocyanurate)

FIGURE 10.13 Formation of polyurethanes and polyisocyanurates (addition polymers). R and R' can be various organic radicals. Partial substitution of a trifunctional isocyanurate will permit cross-linking.

Condensation Polymers

Figure 10.14 shows some examples of condensation polymers. These polymers involve combinations of organic acids (—COOH) with alcohols (—OH), or organic acids with amines (—NH$_2$), with elimination of H$_2$O. Each monomer must contain two functional groups in order for a chain to be constructed. Cellulose is a condensation polymer. It should be noted that Nomex (referred to in Figure 10.14) has a greater thermal stability than nylon, apparently because of the aromatic carbon rings in the polymer chain.

Figures 10.11, 10.12, and 10.14 do not include certain thermosetting polymers. Thermosetting polymers are extensively cross-linked and have complex structures; they do not melt, but char when heated. Examples include urea (NH$_2$CONH$_2$)-formaldehyde (HCHO) polymers such as Plaskon®, phenol (C$_6$H$_5$OH)-formaldehyde polymers such as Bakelite®, and melamine [C$_3$N$_3$(NH$_2$)$_3$]-formaldehyde polymers, such as Formica®.

Many additional polymers exist. Each has a long list of properties. Information about specific polymers can be found in the *Kirk-Othmer Encyclopedia of Chemical Technology.* [9]

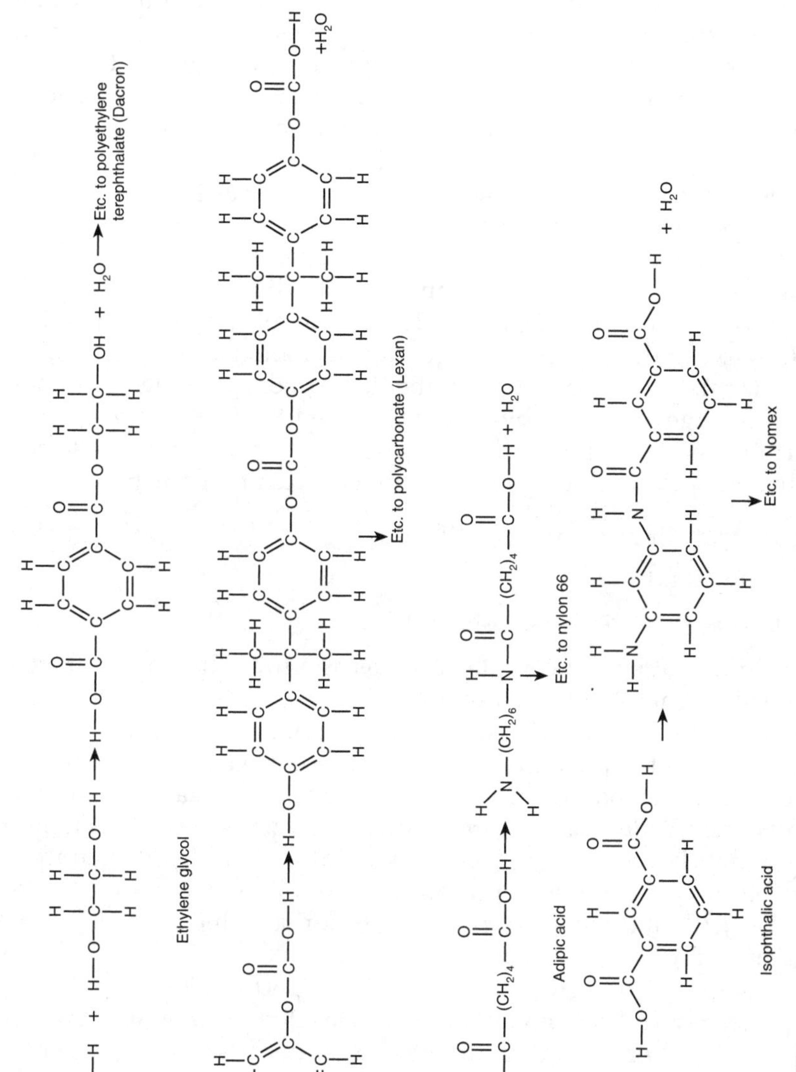

Figure 10.14 Some reactions leading to condensation polymers.

It is important to realize that commercial plastics are not pure polymers, but often contain substantial proportions of additives in order to obtain the desired mechanical, thermal, electrical, and optical properties, or to reduce cost by using inexpensive fillers. Therefore, aside from presence or absence of fire retardants, the properties of a commercial plastic are not identical with those of the polymer from which the plastic is made. Also, polymers can be made into fibers and fabrics, elastomers (rubbery materials), foams (flexible or rigid), coatings, films, or solid plastics. Each of these forms has different fire properties.

A simple way of judging the flammability of a solid is to subject it to the *oxygen index* (OI) test (ASTM D 2863). [10] This test involves igniting a vertical pencil-sized sample at the top, in an oxygen-nitrogen mixture. The proportion of oxygen in the mixture is reduced until the flame goes out. When extinction occurs, the percentage of oxygen by volume is called the *limiting oxygen index* (LOI). Unfortunately, the results of the oxygen index test do not correlate consistently with any of the following important measures of flammability:

1. Ignition delay time versus radiative flux
2. Flame-spread rate
3. Heat-release rate versus radiative flux

It is not surprising that such a complex process as a fire cannot be characterized fully by a single number from a small-scale test.

Nonetheless, the LOI, which is available for many materials, gives a rough idea of relative flammability. Table 10.3 presents some LOI values for synthetic polymers and natural materials. Normal air contains 21 percent oxygen by volume. Under the conditions of the oxygen index test, many of the samples do not burn in air (e.g., red oak). However, if the oxygen index test were to be modified by surrounding the sample either with a heated atmosphere or with radiant heaters, the measured value of LOI would decrease dramatically.

The LOI data show that cross-linked materials perform better in the test than linear polymers, and the more highly halogenated a material is, the better it performs. Addition of fire retardants, to be discussed next, generally increases the LOI.

FIRE RETARDANTS

Both natural and synthetic organic materials often are modified by the addition of chemicals called *fire retardants*. For example, the commonly used term *FR polyurethane* designates fire-retarded polyurethane. The retardants are intended to make the material ignite less readily or burn more slowly, once ignited. Certain additives prevent afterglow of wood. No fire retardant used

TABLE 10.3 Oxygen index values for selected solids

Material	Limiting oxygen index
Polyoxymethylene (Delrin)	15.7
Polyurethane foam (flexible)	16.5
Natural rubber foam	17.2
Polymethyl methacrylate (Plexiglas, Lucite®)	17.3
Polyethylene	17.4
Polypropylene	17.4
Polystyrene	17.6–18.8
Polyacrylonitrile	18.0
ABS	18.3–18.8
Cellulose	19.0
Nylon	20.1–26.0
Wood (birch)	20.5
Polycarbonate (Lexan®)	22.5–28.0
Wood (red oak)	23.0
Wool	23.8
Neoprene (C_4H_5Cl)$_n$	26.3–40.0
Nomex	26.7–28.5
Modacrylic fibers	26.8
Leather	34.8
Phenol-formaldehyde resin	35.0
Polyvinyl chloride	37.1–49
Polyvinylidene chloride	60
Polytetrafluroethylene (Teflon)	95

Source: Values taken from *Fire Protection Handbook*, Table A-5. [11]

as a minor ingredient will prevent combustion completely, especially when a high level of radiant heat flux or a high oxygen concentration is present.

The mechanisms by which fire retardant additives act to modify ignition and burning of solids are imperfectly understood. There are at least four different mechanisms, listed below, and, very often, two or more of these mechanisms might be acting at once.

1. The additive promotes the formation of char and reduces the formation of flammable gases. For example, a solid containing carbon, hydrogen, and oxygen could theoretically decompose to form:

    (a) Char and water vapor, or

(b) Flammable gases such as carbon monoxide, hydrogen, and hydro-carbon gases

A successful additive would promote the first mode of decomposition.

2. The additive releases gases that slow or extinguish the gaseous combustion reactions by:

(a) Dilution and cooling, or

(b) Chemical inhibition of chain reactions.

Such additives often involve halogens.

3. The additive decomposes endothermically, absorbing heat that otherwise would have been available to decompose the base material. For example, hydrated alumina ($Al_2O_3 \cdot 3H_2O$) or limestone ($CaCO_3$) might be mixed with a plastic. When heated, the compound would decompose with absorption of heat and release of an inert gas, either H_2O or CO_2, which cools and dilutes the flame:

$$(AL_2O_3 \bullet 3\ H_2O)_{(s)} \rightarrow (Al_2O_3)_{(s)} + 3(H_2O)_{(g)}$$

$$\Delta H = +162\ kJ$$

$$(CaCO_3)_{(s)} \rightarrow (CaO)_{(s)} + (CO_2)_{(g)};\ \ \Delta H = +178\ kJ$$

4. The additive forms a barrier over the surface, such as a glaze or a foam, which to some degree isolates the subsurface material from the flame above.

The use of fire retardants can pose potential problems. Often, they cause the combustion products to be smokier, more corrosive, or more toxic. Water can wash away certain fire retardants. Fire retardants in structural wood can reduce its strength, and they can interfere with glue or paint on wood. Desirable properties of plastics (mechanical, electrical, thermal, or optical) can be compromised by introducing fire retardants. Some examples of fire retardants are presented below.

When wood is pyrolyzed, the "tar to char ratio" is affected strongly by impregnation by inorganic salts. The positive ions of the most effective salts are ammonium $(NH_4)^+$, sodium, potassium, and zinc, while the negative ions are phosphate $(PO_4)^{-3}$, borate $(BO_2)^{-1}$, silicate $(SiO_3)^{-2}$, sulfate $(SO_4)^{-2}$, and sulfamate $(NH_2SO_3)^{-1}$. Organic compounds containing phosphorus, boron, halogens, or nitrogen (usually as NH_2 compounds) also are used as impregnants for wood.

Instead of impregnating wood, wood is sometimes coated with fire-retardant paints, including *intumescent* coatings, which expand into a foam when heated. These fire-retardant coatings, however, often do not survive long exposure to wet or very humid atmospheres. Other cellulosic materials, such as cotton or paper, also can be treated with fire retardants. An example of a fire retardant used with cotton is tetrakis (hydroxymethyl) phosphonium chloride: $(HOCH_2)_4PCl$.

The fire properties of synthetic polymers can be modified in one of two ways:

1. By copolymerizing them with small proportions of certain monomers, such as vinyl bromide (CH_2=CHBr) or vinyl esters of phosphoric acid, or
2. By mixing nonreactive additives, such as hydrated alumina, into the monomer before polymerizing.

One method widely used for fire-retarding vinyl polymers is to add a combination of a halogenated organic compound and antimony oxide (Sb_2O_3). It is suspected that the effectiveness of this combination is due, at least partly, to formation of a volatile antimony halide or antimony oxyhalide (e.g., SbOCl), which inhibits the gaseous combustion reactions. Note, however, that such fire-retarded systems still can burn readily in a fully developed fire.

Organic phosphorus compounds, such as tricresyl phosphate [$(CH_3C_6H_4O)_3P$=O] or tris (2,3−dibromopropyl) phosphate [$(C_3H_5Br_2O)_3P$=O], are used frequently (instead of more flammable liquids) as plasticizers for polymers. Polyurethane foams generally are fire-retarded by incorporating *halogen* (chlorine or bromine) or phosphorus compounds or both into the materials. Lyons [12] provides further information on the chemistry and uses of fire retardants. Table 10.4 shows the relative use of the various classes of fire retardants in the U.S.

COMPOSITE MATERIALS AND FURNISHINGS

Commonly encountered combustible items are very often composite in nature, rather than consisting of a single chemical substance. For example, an ordinary upholstered chair might consist of a wood frame, a polyurethane

TABLE 10.4 U.S. fire retardants market in 1995 (based on dollar value)

Fire retardant	% of market share
Bromine based	32
Antimony oxide	20
Chlorine based	17
Phosphorus based	17
Alumina trihydrate	11
Other	3

Source: Chemical & Engineering News. [13]

foam pad, cellulosic webbing and welt cords, a neoprene interliner, and a wool cover fabric. Furthermore, the geometry is complex; the most vulnerable point on a chair for ignition by a cigarette is the crevice formed by the juncture of the seat cushion with the armrests or the back.

The complex nature of materials is not limited to furniture. For example, the paneling used for the interior walls of many commercial aircraft consists of acrylic ink-printed polyvinyl fluoride film on fiberglass-epoxy sheets adhered to Nomex honeycomb, which in turn is adhered to a fiberglass-epoxy panel. Electric cables (copper) generally are protected with an outer sheath, often of plasticized polyvinyl chloride, covering an inner insulating layer, perhaps of polyethylene, with or without fire retardant.

It is difficult to predict the combustion characteristics of composite materials simply from a knowledge of their constituents. To the degree that it is practical, combustion characteristics can be determined by conducting fire tests with the actual materials and realistically simulating an ignition source and radiative environment.

Fortunately, ignition studies sometimes can be made with a small representative portion of an item. For example, Figure 10.15 shows how small segments of upholstered foam pads can be used to determine the likelihood that a piece of upholstered furniture will ignite from a lit cigarette. [14] The yields of smoke particles and toxic chemical species are frequently measured by burning small samples of composite materials, generally with an imposed heat flux. [15]

However, the rate of heat release of a complex item like a piece of upholstered furniture, including the period of fire growth, the period of near-maximum heat-release rate, and the burnout period, cannot be determined accurately from small-scale tests. A small-scale test can provide values for maximum heat-release rate versus imposed radiative heat flux, and such values are useful for judging comparative flammability of alternate materials or for making rough estimates of peak heat-release rate of a full-size item (which will be described later).

Large *calorimeters* have been built to measure the heat-release rate *versus* time of individual items of furniture or other combustible assemblies. Figure 10.16 shows such a calorimeter, called a fire products collector, which is capable of handling fires of intensities up to about 10 MW. The fire products (and air) are drawn up into a collector at 28 m^3/s (cold-flow rate). The heat-release rate can be deduced from continuous chemical analysis of a sample drawn from the duct after mixing has occurred, along with a measurement of the flow rate in the duct. This procedure requires a calculation of oxygen consumption by the fire, which causes the O_2/N_2 molar ratio in the exhaust gases to be less than 0.21/0.79. For most combustibles, or for mixtures of combustibles, there are about 13 kJ of heat released per gram of oxygen consumed. [15, 16] This rule is valid even if incomplete combustion occurs.

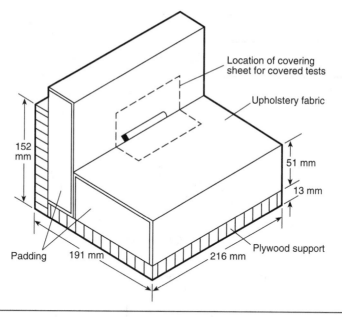

FIGURE 10.15 Mockup design for cigarette ignition test of upholstered chair.

Another way to use the calorimeter is to measure the flow rate and excess temperature of the gas in the duct (relative to ambient temperature) and then calculate the convective component of the heat-release rate. Simultaneously, measurements can be made of the radiative component of the heat-release rate with one or more radiometers viewing the burning object from the side. (This procedure requires an assumption that the radiation is isotropic; that is, it spreads equally in all directions.)

Before the development of fire calorimeters, heat-release rate was estimated by weighing the burning item during the test and multiplying the weight-loss rate by the heat of combustion of the material. This technique does not work well for composite materials. For example, the heat of combustion of polyethylene per gram is 2.5 times as great as that of paper, so if a polyethylene item is in a corrugated paper box, the relative proportions of polyethylene and paper burning at any instant are not known and, therefore, the data cannot be interpreted properly. Furthermore, the weight-loss rate gives no information on the portion of the combustible that vaporizes but fails to burn completely. Finally, if the combustible includes an element that burns to form a solid oxide (e.g., aluminum, boron, silicon), then a weight gain rather than a weight loss might be observed, which would confuse the interpretation of the data.

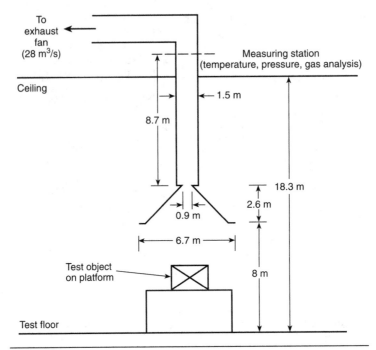

To
exhaust
fan
(28 m³/s)

Measuring station
(temperature, pressure, gas analysis)

Ceiling

8.7 m

1.5 m

18.3 m

2.6 m

0.9 m

6.7 m

Test object
on platform

8 m

Test floor

FIGURE 10.16 Fire products collector (calorimeter) built at Factory Mutual Research Corp.

Table 10.5 shows peak rates of heat release for a variety of full-size objects burned under calorimeters such as described above, as well as some indication of the duration of intense burning. An upholstered sofa might have a peak rate of several megawatts and might burn intensely for several minutes before the rate tapers off to a low value. Additional data on heat release rates are presented by Babrauskas. [17]

One reason for wanting to know the peak rates of heat release of objects is to help predict whether flashover will occur in a room containing one burning object and other not-yet-ignited objects. Flashover, which is the transition to burning of all exposed flammable objects in a room (often characterized by flames projecting out of a door or window), can occur quite suddenly and generally happens when the hot smoke layer under the ceiling reaches about 600°C. At this moment, the radiative heat flux in the room is so great (about 20 kW/m²) that everything flammable ignites.

The minimum rate of heat release needed to cause flashover in a room has been studied extensively. The minimum rate increases with the size of the room, and depends, in a complicated way, on the ventilation in the room. If there is too little ventilation, flashover cannot occur. If there is an excessive

TABLE 10.5 Heat-release rates for various combustible items

Item	Peak rate of heat release (MW)	Time of intense burning (s)
Wood pallet stack:		
1.22 m high	3.6	460 (>2.5 MW)
3 m high	7.9	
6 m high	14	
Upholstered furniture (various)	0.3–3.1	
Upholstered sofa	3	200 (>1 MW)
Upholstered chair	1.9	130 (>0.5 MW)
Mattress, twin-size (various)	0.1–2.7	
Latex® foam pillow	0.12	240 (>0.05 MW)
Polyurethane pillow	0.04	
Feather pillow	0.02	270 (>0.01 MW)
Television set	0.23, 0.29	600 (>0.1 MW)
Waste basket, 6.6 L, containing empty milk cartons	0.07	200 (>0.05 MW)
Trash bags containing paper (3 bags, 3.51 kg total)	0.35	200 (>0.15 MW)

Source: Babrauskas. [17]

amount of ventilation, the excess air flow dilutes and cools the smoke, so a larger rate of heat release is needed to reach the critical temperature condition for flashover. The materials of construction and thickness of the ceiling and upper walls are also important factors in determining whether flashover will occur, and, if so, how soon. Drysdale [18] gives quantitative relationships that can be used to calculate the critical fire size for flashover.

For example, to cause flashover in an average-size bedroom lined with plasterboard, with an open door, a heat-release rate of about 1 MW that continues for about 5 min would be needed. If the ceiling had been bare concrete instead of plasterboard, a larger fire or a longer duration of fire would be needed to cause flashover, because the ceiling would heat up much more slowly. The time to reach flashover is less if the rate of heat release is larger than the threshold value. (*See Figure 10.17.*) The changing conditions prior to flashover are affected by so many variables that computer models are needed to predict flashover in particular cases [19] as described in Chapter 13.

Babrauskas and Krasny [20] described a formula that approximately correlates the peak heat-release rate, q_p (in kW), of an item of upholstered furniture, as measured in a large calorimeter, with a small-scale measurement of a

FIGURE 10.17 Flashover. In less than 2 $\frac{1}{2}$ minutes after ignition, fire in this test room was pushing heavy flames out the doorway. Courtesy National Institute of Standards and Technology.

sample of the same material. The small-scale measurement, denoted q_s, is the average heat-release rate (in kW/m²), over a 180-s period after ignition, measured with an imposed radiative flux of 25 kW/m² on the sample. The other variables are m, the combustible mass of the furniture item in kilograms; s, the style factor, equal to 1.0 for "plain" construction and 1.5 for "convoluted" shapes; and f, the frame factor, equal to 0.2 for charring plastic frames, 0.3 for wood frames, 0.6 for melting plastic frames, and 1.7 for noncombustible frames. The correlating formula is

$$\frac{q_p}{q_s} = 0.63 \ m \bullet s \bullet f$$

Note that this formula implies a direct proportionality between the peak burning rate and the mass of the object, and also suggests that burn-through or melting of the frame, causing collapse, results in a lower peak burning rate than with a noncollapsing metal frame.

METALS

The oxidation of a metal usually is a highly exothermic process. For example, the heat of oxidation of aluminum is 31 kJ/g, which is nearly twice that of wood. On the other hand, the thermal conductivity of aluminum is about 1000 times that of wood, so it is virtually impossible to heat the surface of a massive piece of aluminum to a high temperature without at the same time heating the entire mass to nearly the same temperature. Thermal conductivities (and electrical conductivities) of metals, in general, are many times as great as the conductivities of wood or synthetic polymers.

Accordingly, massive pieces of metal are not expected to burn, except after being heated for a long period. However, if a metal is in the form of a fine powder, suspended in air, the ignition source could heat the nearby small particles, and the thermal conductivity within each particle would become irrelevant. Indeed, most metal powders are combustible. In certain cases, the metal does not have to be as fine as a powder to burn, but could be in the form of chips or shavings, as from machining.

Solid or liquid organic substances, with the exception of carbon, burn in the vapor phase. The reason that carbon burns on its surface and not in the gas phase is that a temperature of about 3900°C is needed to vaporize it. Such a temperature is much higher than would be present in an air-supported flame. Metals can be divided into two categories:

1. Those, like carbon, that burn on their surface
2. Those that burn in the vapor phase

The first four metals listed in Table 10.6, which are believed to burn on their surface, all have extremely high boiling points (BP), which are higher than their flame temperatures.

The flame temperature of a metal is limited by the boiling point of its oxide, because the oxides have a very high heat of vaporization, and the flames generate only enough heat for partial vaporization of the oxides. As Glassman [21] noted, the boiling points of the oxides of the first four metals in Table 10.6 are below the boiling points of the metals. The remaining six metals listed in Table 10.6 vaporize while burning, and the boiling points of the oxides are above the boiling points of the metals. Note that boron and silicon, which are included in Table 10.6, are not exactly metals. These elements are intermediate in their properties between metals and nonmetals.

As a general rule, materials that burn on the surface will burn more slowly than materials that can vaporize first, because the vapors can more readily contact the surrounding oxygen in the air. Another general characteristic of metals, especially hot metals, is that most of them can react rapidly and exothermically with water to form hydrogen. For example,

$$2\,Al + 3\,H_2O_{liq} \rightarrow Al_2O_3 + 3\,H_2 + 819\,kJ$$

Table 10.6 Metals that burn on their surface and metals that burn in the vapor phase

Metal	Burning mode	Metal BP (K)	Metal oxide BP (K)
Silicon	Surface	3514	~2500
Titanium	Surface	3591	~3300
Boron	Surface	3931	2316
Zirconium	Surface	4777	4548
Potassium	Vapor	1037	1154
Sodium	Vapor	1156	2223
Magnesium	Vapor	1378	3533
Lithium	Vapor	1620	2836
Calcium	Vapor	1767	~3800
Aluminum	Vapor	2768	~3800

Consequently, if a partially wet metal is burning, a hydrogen fire must be dealt with as well as a metal fire. Also, water under hot molten metal will turn rapidly into steam and erupt, throwing molten metal around. Some details about the combustion of specific metals follow.

Magnesium

Magnesium in the form of powder, ribbons, or shavings can ignite under some conditions at about 500°C (932°F). However, massive magnesium must be heated to its melting point (650°C; 1202°F) for ignition to occur. Some magnesium alloys have lower ignition temperatures. Magnesium chips wet with animal or vegetable oils have been known to ignite spontaneously. Molten magnesium in contact with iron oxide (rusted iron) produces a highly energetic *thermite-type* reaction, that is,

$$3 \text{ Mg} + \text{Fe}_2\text{O}_3 \rightarrow 3 \text{ MgO} + 2 \text{ Fe} + 971 \text{ kJ}$$

When finely divided, magnesium will burn in an atmosphere of any of the following pure gases, to yield the product shown:

Steam: $\text{MgO} + \text{H}_2$

Carbon dioxide: $\text{MgO} + \text{CO}$

Nitrogen: Mg_3N_2

Halon 1301 (CF_3Br): MgF_2, MgFBr, MgBr_2, C

Magnesium powder mixed with Teflon $(C_2F_4)_n$, even under an inert gas such as helium or argon, will burn to form MgF_2 and carbon. Clearly, the choice of agent to use in fighting a magnesium fire is severely limited. Extinguishing agents are discussed in Chapter 14.

Aluminum

Aluminum has a much higher boiling point than magnesium (2495°C versus 1105°C), so it cannot be ignited as readily. However, aluminum powder, flakes, very fine chips, and shavings can be ignited and, once ignited, will burn like magnesium chips or powder. Aluminum powder is a major ingredient in solid-propellant rocket fuels, and burns rapidly in rockets, to give a very high exhaust temperature.

Aluminum will react with iron oxide (the thermite reaction). However, aluminum will not burn in nitrogen. The relatively low melting point of aluminum (660°C) will favor melting of aluminum siding or roofing in a fire.

Iron and Steel

Iron and steel generally do not burn in air but can burn in pure oxygen. However, fine steel wool or steel dust in air can be ignited with a torch. Pure iron powder, when exposed to air for the first time after manufacture, can ignite spontaneously.

Titanium

Massive pieces of titanium generally cannot be ignited. In spite of its high melting point (1660°C) and high boiling point (3318°C), fine turnings and thin chips of titanium can be ignited with a match. Once ignited, vigorous burning results. Titanium dust clouds will ignite in air when heated. Like magnesium, titanium dust will burn in pure carbon dioxide or pure nitrogen.

Alkali Metals

Lithium, sodium, and potassium have some unusual fire properties. They all have low melting points (186°C, 98°C, and 62°C, respectively). Sodium or potassium, on contact with water at room temperature, will generate hydrogen exothermically and can burst into flame spontaneously. Lithium reacts more slowly with water, without bursting into flame. These *alkali metals* can be ignited by heating in dry air. Once ignited, they burn vigorously, producing dense white clouds of metal oxides. Lithium differs from sodium and potassium in that it will burn in pure nitrogen.

Alkali metals often are stored under kerosene or oil to isolate them from air and moisture. Alkali metals will react violently with halogenated hydrocarbons or sulfuric acid.

NaK is a sodium-potassium alloy with a very low melting point; it is a liquid in the vicinity of room temperature. NaK is used as a heat-transfer fluid; its fire properties are similar to those of sodium and potassium, but its reactions are claimed to be more vigorous.

Other Metals

Discussion of the fire properties of zirconium, hafnium, calcium, zinc, uranium, thorium, and plutonium is given by Tapscott. [22]

Problems

1. Why will a single wooden log, if ignited, soon self-extinguish, while a group of logs near each other will continue burning?

2. What is the difference between spontaneous ignition and piloted ignition of a solid?

3. If a pilot flame is present, how intense must a thermal radiative flux be to ignite wood? How does this flux compare with the intensity of sunlight (about 0.7 kW/m^2 at noon in the tropics)?

4. How does the formation of a char layer affect the burning of a solid?

5. Why is upward flame spread over a vertical surface more rapid than downward flame spread? Could the upward rate ever be as much as 100 times as great as the downward rate?

6. Why do flames spread more rapidly on thin materials?

7. The top of a Plexiglas table is burning. The flame has spread to the edges. Of the heat transferred from the flame to the table surface (to supply the heat of gasification), what fraction is radiative and what fraction is convective?

8. What are the upper and lower extremes of the moisture content of wood?

9. What do wood, paper, and cotton have in common?

10. What is the main chemical difference between cotton and wool?

11. What is the difference between a linear polymer and a cross-linked polymer?

12. What is a monomer? A copolymer?

13. What is the oxygen index test? Why are the results of this test of limited use?

14. What are the four ways in which fire retardants can act?

15. Under what conditions can the presence of fire retardants cause problems?

16. How can a fire products calorimeter be used to characterize the burning of an upholstered chair?

17. What is flashover?

18. What determines whether metals burn on their surface or burn by vaporizing first?

19. Why is it not advisable to apply water to burning metals?

References

1. Butler, C. P., "Notes on Charring Rates in Wood," *Fire Research Note 896,* British Fire Research Station, Borehamwood, Herefordshire, England, 1971.

2. Smith, E. E., "Heat Release Rate of Building Materials," *Ignition, Heat Release and Noncombustibility of Materials,* ASTM Special Technical Publication No. 502, American Society for Testing and Materials, W. Conshohocken, PA, 1972, pp. 119–134.

3. Tewarson, A., and R. F. Pion, "Flammability of Plastics-I, Burning Intensity," *Combustion and Flame,* Vol. 26, 1976, pp. 85–103.

4. Babrauskas, V., "Development of the Cone Calorimeter — A Bench-Scale Heat-Release Rate Apparatus Based on Oxygen Consumption," *Fire and Materials,* Vol. 8, 1984, pp. 81–95.

5. Magee, R. S., and R. D. Reitz, "Extinguishment of Radiation-Augmented Plastics Fires by Water Sprays," *Fifteenth Symposium (International) on Combustion,* The Combustion Institute, Pittsburgh, PA, 1975, pp. 337–347.

6. Tewarson, A., "Generation of Heat and Chemical Compounds in Fires," *The SFPE Handbook of Fire Protection Engineering,* 1st ed., National Fire Protection Association, Quincy, MA, 1988, Section 1, pp. 179–199.

7. de Ris, J., "Fire Radiation — A Review," *Seventeenth Symposium (International) on Combustion,* The Combustion Institute, Pittsburgh, PA, 1979, pp. 1003–1016.

8. *Kirk-Othmer Encyclopedia of Chemical Technology,* 3rd ed., J. Wiley, New York, 1984, Vol. 24, pp. 579–611.

9. *Kirk-Othmer Encyclopedia of Chemical Technology,* 3rd ed., J. Wiley, New York, 1978–84, 24 vols.

10. ASTM D 2863, *Standard Test Method for Measuring the Minimum Oxygen Concentration to Support Candle-Like Combustion of Plastics (Oxygen Index)*, American Society for Testing and Materials, W. Conshohocken, PA., 1991.

11. Cote, A. E., *Fire Protection Handbook*, 18th ed., National Fire Protection Association, Quincy, MA, 1997.

12. Lyons, J. W., *The Chemistry and Uses of Fire Retardants*, Wiley-Interscience, New York, 1970.

13. *Chemical & Engineering News*, American Chemical Society, Washington, DC, Feb. 24, 1997, p. 20.

14. Gann, R. G., R. H. Harris, J. F. Krasny, R. S. Levine, H. Mitler, and T. J. Ohlemiller, "Effect of Cigarette Characteristics on the Ignition of Soft Furnishings," Technical Note No. 1241, National Bureau of Standards, Jan. 1988.

15. Tewarson, A., "Experimental Evaluation of Flammability Parameters of Polymeric Materials," *Flame-Retardant Polymeric Materials*, Vol. 3, Plenum Press, New York, 1982, pp. 97–153.

16. Huggett, C., "Estimation of the Rate of Heat Release by Means of Oxygen Consumption Measurements," *Fire and Materials*, Vol. 4, 1980, pp. 61–65.

17. Babrauskas, V., "Burning Rates," *The SFPE Handbook of Fire Protection Engineering*, 2nd ed., National Fire Protection Association, Quincy, MA, 1995, Section 3, pp. 1–15.

18. Drysdale, D., *An Introduction to Fire Dynamics*, 2nd ed., J. Wiley, New York, 1998.

19. Snell, J. E., "Fire Safety Review," *Fire and Materials*, Vol. 13, 1988, pp. 5–12.

20. Babrauskas, V., "Upholstered Furniture and Mattresses," *Fire Protection Handbook*, 18th ed., National Fire Protection Association, Quincy, MA, 1997, Section 4, p. 190.

21. Glassman, I., *Combustion*, 3rd ed., Academic Press, New York, 1996.

22. Tapscott, R. E., "Metals," *Fire Protection Handbook*, 18th ed., National Fire Protection Association, Quincy, MA, 1997, Section 4, pp. 182–189.

CHAPTER 11

Combustion Products

SMOKE

Fire scientists usually define *smoke* as the mixture of tiny particles and gases produced by a fire. The particles are mainly of two kinds:

1. Carbonaceous solid particles (soot), producing black smoke

2. Liquid droplets (aerosol mist), which form as some gas molecules cool and condense, producing light-colored smoke.

(See Figure 11.1.) In some cases, the particles might also include mineral matter that was present originally in the combustible.

Two difficulties are raised if smoke is defined as only tiny particles, and not the accompanying gases. First, the toxic effects of the fire products are very important. The mixture of particles and gases sometimes produces synergistic effects; because of interactions, the whole can be different from the sum of its parts insofar as toxicity is concerned. Second, the tiny liquid droplets that form as the fire gases cool increase in size as further cooling occurs because of condensation on the droplet surfaces. Therefore, the relative proportions of particles and gases depend strongly on the temperature.

The formation of soot particles in flames and the ability of hot smoke to emit thermal radiation were discussed in Chapter 8. The toxicity of smoke is discussed in the next section of this chapter, Toxicity of Fire Products. In addition to toxicity, the vision-obscuring aspect of smoke is important, in that it

FIGURE 11.1 Two basic kinds of smoke that emanate from fires.

can prevent timely escape or rescue. From a positive viewpoint, smoke detectors provide a convenient way of detecting a fire.

The type and quantity of smoke produced in any fire depends not only on what material is burning but also on the burning conditions. For example, the smoke from a flaming fire is different from that of a smoldering (nonflaming) fire of the same material. A fire in one region of a compartment can produce radiant heat that causes a material in another region of the compartment to pyrolyze without igniting, producing vapors that later cool and condense into an aerosol type of smoke. For underventilated flaming fires, the smoke yield can be either greater or less than for fully ventilated fires of the same material, depending on conditions.

Measurement Procedures for Smoke Particulates

There are two principal methods for quantifying the particulate content of smoke produced in an experimental fire:

1. Collecting and weighing the particulates and then calculating the ratio to the weight loss of the combustible
2. Measuring the attenuation of a beam of light passing through the smoke and normalizing the result

The second method is used more often because it is more convenient and gives information directly applicable to the question of visibility through smoke. The two methods roughly correlate with one another, but they are not directly proportional. *(See Figure 11.2.)*

If the smoke particles are collected, the particle-size distribution can be measured microscopically. If the smoke is measured by passing light

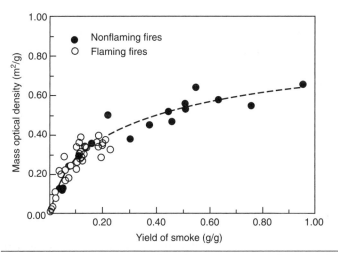

FIGURE 11.2 Comparison of smoke measurements by optical technique at a wavelength of 0.633 μm and by weighing the collected smoke. Source: Tewarson. [1]

through it, the mean particle size can be calculated by measuring the different optical responses of the particles to light of different wavelengths. Some idea of particle-size distribution also can be obtained by this method, using light of three wavelengths. This procedure requires making assumptions about particle shape and optical properties. Measurements by both these procedures show that smoke particles produced in fires range from 0.3 μm to 1.6 μm in mean diameter, depending on combustion conditions, chemical nature of the combustible, and changes occurring as the smoke ages. [2]

The most common optical method for measuring smoke is to project a beam of light through the smoke, across a gap x meters wide between a light source and a photometer. The light beam leaving the source has an intensity, I_o, while the light arriving at the photometer has a lower intensity, I, because some of the light has been scattered either sideways or backward, or the light has been absorbed because of interactions with the particles. The relationships between I, I_o, and x are given by *Bouguer's law*, which also is known as the *Beer–Lambert law*:

$$\frac{I}{I_o} = e^{-kx} = 10^{-kx/2.303}$$

where k (reciprocal meters; called the *extinction coefficient*) is a measure of the opacity of the smoke. This variable depends on the number of smoke particles per unit volume, and also on the size, shape, and refractive index of these particles, as well as the wavelength of the light used. The term $kx/2.303$ usually is defined as the *optical density*, D, of the smoke. (However, Drysdale [3] defined the optical density, D, as $10 \cdot kx/2.303$.)

Assume that, in a given fire, m grams of combustible vaporize, followed by incomplete combustion, and the resulting smoke is contained in a volume of V cubic meters. (Or, assume that the combustible is vaporizing at the rate of m grams per second, and the smoke is flowing away at the rate of V cubic meters per second.) The optical density, D, of the smoke would be expected to be directly proportional to m, for a given V, and inversely proportional to V, for a given m. Hence, the optical density should be proportional to m/V. Also, because D is defined as $kx/2.303$, D is proportional to x. Accordingly, if a new quantity, D_m, the *mass optical density*, is defined as

$$D_m = \frac{(D/x)}{(m/V)} = \frac{VD}{mx}$$

then D_m is a normalized value of optical density, which should be independent of the values of m, V, and x used in the measurement, and dependent only on the nature of the combustible and the way it is burned (pyrolyzing, smoldering, flaming, partially ventilated, fully ventilated, high radiation level, or low radiation level).

Quantity of Smoke Particles Produced

Tables 11.1 and 11.2 show measurements of the smoke particles produced from selected gaseous, liquid, and solid combustibles burned under laboratory conditions. Additional data of this type are available. [1, 2, 3, 4] Some results are expressed in terms of grams of particulate collected per gram of combustible vaporized, and other results are expressed as mass optical

TABLE 11.1 Smoke particles produced from small-scale burning of various liquids and gases under well-ventilated conditions

Material	Smoke yield (g/g)	Mass optical density D_m (m²/g)
Methanol	<0.001	—
Ethane	0.008	0.024
Ethanol	0.012	—
Acetone	0.014	—
Propane	0.006–0.025*	0.02–0.08*
n-Octane	0.039	0.20
Ethylene	0.045	0.20
Acetylene	0.13	0.32
Benzene	0.18	0.36
Styrene	0.18	0.35

*Depends on flow rate to burner.
Source: Tewarson, 86–191. [1]

TABLE 11.2 Smoke particles produced from small-scale flaming, smoldering, or pyrolyzing of various solids under well-ventilated conditions

Material	Smoke yield (Flaming: g/g)	Mass optical density D_m (m²/g)	
		Flaming	Smoldering or pyrolyzing
Pine[1]		0.038	
Sugar pine[2]	0.004	0.013	
Red oak[2]	0.002–0.011	0.01–0.04	
Red oak [1]	0.015	0.037	0.33
Plywood[3]			0.29
Plywood[4]		0.017	0.17
Plexiglas[1]	0.022	0.11	
Plexiglas[2]	0.012–0.016	0.07	
Plexiglas[3]			0.15
Delrin[1]	<0.001	0.01	
Nylon[1]	0.075	0.23	
Neoprene[3]		0.40	0.55
PVC, rigid[4]		0.17	0.18
PVC, rigid[3]		0.34	0.12
PVC, rigid[1]	0.17	0.40	
PVC, plasticized[3]			0.64
Polystyrene[3]		0.79–1.4	
Polystyrene [1]	0.16	0.34	
Polystyrene foam[3]		0.79	
Polystyrene foam[1]	0.18–0.21	0.34–0.37	
Polyethylene[1]	0.06	0.23	
Polyethylene[3]		0.29	
Flexible polyurethane foams[1]	0.13–0.23	0.29–0.34	0.30-0.56

[1]Tewarson, pp. 186–191. [1]   [3]Mulholland, Section 2, p. 223. [2]
[2]Mulholland et al., p. 352. [4]   [4]Drysdale, p. 360. [3]

density, D_m, which is defined above. In each case, the laboratory "fire" was reasonably well ventilated.

For a number of entries in Table 11.2, measurements of the same or similar materials were available from several sources; the results usually do not agree very closely. This inconsistency strongly suggests that the results are sensitive to the exact details of the apparatus and experimental procedure. The results also depend on the degree of ventilation. Precise predictions of smoke yields from fires are very difficult to make.

Still, Tables 11.1 and 11.2 do show very large differences from one material to another. Methanol burns with practically no particulate production. Wood, under flaming combustion conditions, produces approximately only 1 percent of smoke particles relative to weight loss, but, under smoldering or pyrolyzing conditions, produces approximately 10 percent of smoke particles. Some plastics, such as polystyrene, can produce up to 20 percent of smoke particles relative to weight loss under flaming conditions, sometimes even more under smoldering or pyrolyzing conditions. On the other hand, other plastics, such as Plexiglas, produce only 1 to 2 percent of smoke particles, and Delrin produces practically no particles under flaming conditions. Accordingly, while the numbers measured in the laboratory do not strictly predict smoke character in a real fire, a comparison of materials is, nevertheless, worthwhile in obtaining a relative idea of smoke particle production in a fire.

The molecules of methanol (CH_3OH), Delrin ($CH_2O)_n$, Plexiglas ($C_5H_8O_2)_n$, and cellulose ($C_6H_{10}O_5)_n$, which do not produce many smoke particles in flaming combustion, all contain substantial oxygen. On the other hand, molecules containing halogen atoms X (= F, Cl, or Br) tend to burn with high smoke-particle production, presumably because HX splits out of the decomposing molecule, leaving an unsaturated molecule that is transformed readily into a cyclic hydrocarbon and then to soot. (See Chapter 8.) Also, the presence of HX and H_2O in the combustion products can lead to the formation of an acid mist (aerosol). Of course, any combustible having an aromatic (aryl) structure, such as benzene or styrene, is converted very readily to soot in a flame.

When Babrauskas and coworkers [5] performed radiation-assisted flaming-combustion tests, they reported an 86 percent increase in mass optical density when 12 percent of a brominated fire retardant and 4 percent of antimony oxide were incorporated into a polystyrene material. Such effects have been noted frequently by others, and there has been much discussion about the trade-off between reduced combustibility and increased smoke yield when halogenated fire retardants are used.

The question remains as to how accurately these laboratory measurements of smoke-particle production can be correlated with smoke generated in real fires. Several studies have been made in which small-scale and large-scale fire data have been compared, with limited success. [5, 6] The discrepancies probably can be attributed to two main factors:

1. In a realistic fire in a compartment, a combination of flaming combustion and pyrolysis (decomposition without flaming) is often occurring at the same time.

2. The restricted ventilation and partial recirculation of combustion products to the flame in a compartment fire is difficult to simulate in a bench-scale test.

Visibility Through Smoke

Visibility through smoke is often expressed as the distance over which an exit sign can be seen. A sign that emits light is easier to see than one that must be viewed in reflected light. Mulholland [2] provides a means for relating the distance d (in meters) over which the sign can be seen to the mass optical density, D_m (in square meters per gram), of the smoke:

$$d = \frac{1.3}{D_m(m/V)}$$

for a light-reflecting sign, and

$$d = \frac{3.5}{D_m(m/V)}$$

for a light-emitting sign.

In these formulas, m/V is the mass loss (in grams) of the combustible divided by the volume (in cubic meters) that contains a uniform mixture of the smoke. Such formulas are not precise because of the variability of human vision and the effect of the varying background illumination level. Visibility of light-emitting exit signs in a smoke-filled room is improved significantly if the ambient lighting is at a fairly low level, because higher levels of ambient light obscure signs due to light-scattering. Another important variable is the frequent presence of eye irritants, such as acrolein or hydrogen chloride in the smoke.

Visible light has wavelengths ranging from 0.4 µm (blue light) to 0.8 µm (red light). When light encounters a smoke particle whose diameter is comparable to the wavelength of the light, then the light will be scattered at various angles from its original direction, including back-scattering. This scattering effect causes objects viewed through smoke to appear blurry or indistinguishable. *(See Figure 11.3.)*

A given smoke sample will contain particles over a range of diameters, but generally, a large mass fraction of the particles will be between 0.1 and 1 µm (close to the wavelengths of visible light). Accordingly, scattering by smoke has an important effect on visibility during a fire.

Objects can emit infrared light as well as emitting or reflecting visible light. Infrared light can have much longer wavelengths than visible light, including wavelengths that are large compared with the diameter of smoke particles. Hence, there is much less scattering. If an observer were equipped with an infrared-sensitive video camera, which operated to wavelengths as long as 14 µm, flame through smoke could be seen, as well as an opening at the end of a smoke-filled corridor, for example, that would otherwise be invisible. Such video cameras, which are now available, also allow a rescuer to see warm bodies of unconscious fire victims in smoke-filled rooms.

FIGURE 11.3 The scattering effect of smoke. The ability to see people or light sources is diminished considerably by the scattering effect of smoke.

Smoke Detectors

The two main classes of smoke detectors follow:

1. Those based on light scattering by smoke particles (optical detectors)
2. Those based on the interception of gaseous ions and electrons by smoke particles (ionization detectors)

Both types of detectors can be designed to be exceedingly sensitive. However, to avoid frequent false alarms caused by dust; cigarette smoke; cooking fumes; and aerosols, such as hairsprays or furniture polish, the detectors must be designed so that they respond only to fairly high concentrations of particles. Optical detectors are designed to respond to a smoke obscuration level of a little less than 13 percent per meter for gray (cellulosic) smoke.

Both types of smoke detectors require a source of electricity, often a battery. In the most common type of optical detector, a light-emitting diode emits a beam that passes through a volume of smoke that has flowed into the detector. A photoelectric cell measures the fraction of this light that is scattered in a particular direction. Another type of optical detector projects a beam of light over a long distance, and the attenuation of the beam is used as a measure of the presence of smoke particles.

The ionization smoke detector contains a tiny quantity of a radioactive element, americium, which creates ions and free electrons in the air passing through it. The ionization current through this air is measured continuously. If smoke particles are present, some of the ions and electrons will attach themselves to particles, which leads to a reduction in charge mobility and consequently a reduction of ionization current.

For a given number of particles per unit volume, for either the light-scattering type or the ionization type of detector, the larger the particle size, the greater is the signal. However, for this given number of particles per unit volume, the signal is proportional to the first power of the particle diameter for the ionization detector, while the signal from the light-scattering detector is proportional to a much higher power of particle diameter (except for very large particles). Accordingly, if both types of detectors are adjusted to have the same sensitivity for a certain size of particle, then the ionization detector will be more sensitive for particles smaller than this size, while the optical detector will be more sensitive for particles larger than this size.

It so happens that, for wood or paper, flaming fires produce small particles (fine soot), while smoldering fires produce larger particles (aerosol droplets). Therefore, an ionization detector will respond sooner for a flaming cellulosic fire, and an optical detector will respond sooner for a smoldering fire. For plastics fires, no such simple rules apply.

It is important to recognize that, no matter which type of detector is used, smoke can be detected well before the concentration at the detector has reached levels where visibility is seriously impaired or short-term toxic concentrations have been reached.

TOXICITY OF FIRE PRODUCTS

It is widely recognized that fire victims die because of smoke inhalation more often than from burns or from building collapse. Statistics from the National Academy of Sciences [7] show that about 6000 fire deaths occur in the United States each year, and 70 to 80 percent of these are caused by the physiological effects of smoke (defined as a mixture of gases and particles).

Carbon monoxide is a widely known toxicant in smoke, but do other toxic species in smoke play an important role in fire deaths? A complete answer to this question is not available because experimentation with humans is unacceptable. However, current knowledge gives a partial answer.

There is no doubt that carbon monoxide is the principal toxicant in many cases. In the state of Maryland, autopsies were performed on almost all fire death victims over a period of several years, as part of a Johns Hopkins University research program. [8] Of the 530 victims, 60 percent were found to have more than one-half of the hemoglobin in their blood converted to carboxyhemoglobin, clearly as a consequence of having inhaled carbon monoxide shortly before dying. From experiments in which test animals are exposed to carbon monoxide, it has been established that conversion of more than about one-half their hemoglobin to carboxyhemoglobin causes incapacitation and usually death.

In the Maryland study, an additional 20 percent of the fire victims had a somewhat lower carboxyhemoglobin level, but were found also to have preexisting cardiovascular disease; this suggests that they were vulnerable to lower exposures of carbon monoxide than healthy people. Thus, it was judged that 80 percent of the victims could have been killed by carbon monoxide. Of the remaining 20 percent, 11 percent died from burns, and the cause of death was not established for 9 percent.

Hydrogen cyanide (HCN) was discovered in the blood of a number of victims in the Maryland study [8] and was suspected to be a cause or contributor to death in those cases. Autopsies performed in a more recent investigation in Pennsylvania [9] confirmed that hydrogen cyanide was the primary cause of death for some of the 136 fire victims studied. Of course, hydrogen cyanide can only be present in smoke if the combustible contains organically bound nitrogen.

If an occupant of a building is confronted with a growing fire, exposure to the smoke might have incapacitating effects that hinder escape capability well before a fatal dose is acquired. These incapacitating effects can influence mobility and/or clear thinking.

There are two ways of assessing the toxicity of smoke from a particular combustible:

1. Perform a detailed chemical analysis of the smoke

2. Expose laboratory animals to the smoke (bio-assay) according to a test procedure such as described by Purser. [10]

Neither method is entirely satisfactory.

The principal criticism of the chemical analysis method is that some unexpected, highly toxic species that could be present in low concentration might not be found. (It is not easy to do a chemical analysis unless one knows what to look for.) A second criticism is that the toxic effect of the complex mixture of gases and particulates in the smoke cannot necessarily be predicted from a knowledge of the toxic effects of the individual components. That is, a mixture of components could be much more toxic than the effect of each species alone.

On the other hand, the bio-assay method can be criticized because the response of test animals to smoke is not necessarily the same as that of humans. Furthermore, the method is expensive and slow because the test animals might have to be observed for weeks after the exposure. A number of groups of test animals must be exposed at different smoke concentration levels to find the condition that will kill approximately half of a group of perhaps six animals (referred to as LC_{50}—the lethal concentration that will kill 50 percent of the animals).

Both the chemical and the bio-assay methods require an arbitrary selection of some standardized way of burning a sample. The constituents in the smoke will be sensitive to the choice of burning technique. Therefore, the test smoke can differ significantly from smoke found in a real fire.

Finally, reference must be made to the concept of dose as it relates to concentration of the toxic species. For example, a dose might consist of exposure to 5000 ppm (parts per million) of a toxicant for 2 minutes, or 500 ppm of the same toxicant for 20 minutes. These two doses are not necessarily equivalent.

For these and other reasons, including the variability of humans, knowledge of smoke effects on humans is far from complete at this time, and is a subject of continuing research.

Carbon Monoxide

When a carbon-containing combustible burns in a compartment with less than the stoichiometric requirement of air available, the carbon monoxide content of the smoke can be as high as 5 or 6 percent by volume. Even with a more adequate air supply, a fire burning with a very sooty, highly luminous flame (e.g., benzene, polystyrene, or polyvinyl chloride) can have 2 percent of carbon monoxide in the smoke (2 percent carbon monoxide is 20,000 ppm).

If a fire compartment is lined with wood on the upper walls or ceiling, and this compartment fills with hot fire gases highly deficient in oxygen, then the wood will pyrolyze, producing very high concentrations of carbon monoxide, up to 14 percent by volume.

Recent research by Gottuk, Roby, and Beyler [11] and by Pitts [12, 13] investigated the levels of carbon monoxide present in fire gases after flashover, in an attempt to provide a quantitative basis for predicting toxic effects of a fire. They concluded that the carbon monoxide level depends on the following factors:

1. "Global equivalence ratio" in the upper part of the compartment, which is the overall ratio of combustible to oxygen, relative to the stoichiometric ratio

2. Temperature of the hot layer

3. Mixing of air into the hot layer

4. Surface area of any wood (e.g., paneling) in contact with the hot layer

In view of these many complexities, as well as the possibility that conditions in the hot layer may not be uniform, it appears that accurate predictions of carbon monoxide level are difficult to make.

According to Hartzell [14], a person exposed to air containing 20,000 ppm CO would be incapacitated in 2 minutes. Death would occur a few minutes later. Carbon monoxide has a strong affinity for the hemoglobin of the blood, which normally carries oxygen from the lungs to the brain and body. As hemoglobin reacts with carbon monoxide it forms carboxyhemoglobin, which can no longer carry oxygen. Table 11.3 shows the symptoms that develop in a healthy person as the percentage of carboxyhemoglobin increases.

TABLE 11.3 Symptoms as the percentage of hemoglobin converted to carboxyhemoglobin increases in the human bloodstream

Percent carboxyhemoglobin	Symptoms	
	Short term	Long term
10	Judgment inefficiencies	
10–20	Labored breathing	
20–30	Headache	
30–40	Nausea, dizziness	
40–50	Fainting	Death for some
50–60	Convulsions	Death for most
>60	Coma	Death for all

Figure 11.4 shows the time required to incapacitate and kill rats who were exposed to various concentrations of carbon monoxide. Less than 0.2 percent CO (<2000 ppm) can incapacitate rats in 20 minutes.

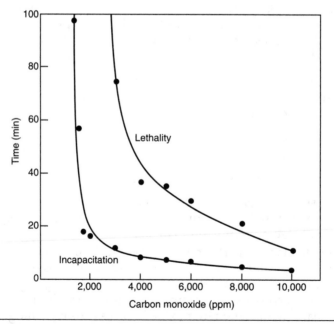

FIGURE 11.4 Carbon monoxide intoxication of rats. From Hartzell et al., p. 1061. [15]

Haber's rule, which is not valid for very low or very high concentrations of toxicants, states that *a given toxic effect is produced by a given product of concentration and time*. For example, 5000 ppm of CO for 10 minutes should cause the same effect as 10,000 ppm for 5 minutes. Hartzell [14] has proposed that, when the carbon monoxide concentration-time product exceeds 35,000 ppm-minute, there is a high probability of incapacitation for many people.

Carbon Dioxide

Carbon dioxide is not highly toxic. Exposure to 10 percent carbon dioxide in air can cause headache and dizziness, but the most immediate effect is to stimulate the rate and depth of breathing (8 to 10 times the normal level). Even 4 percent carbon dioxide is enough to double the lung intake.

The effect of carbon dioxide on breathing rate is important for the following reason. A person exposed to a mixture of 0.5 percent CO and 5 percent CO_2 in air will be overcome sooner than if exposed to 0.5 percent CO in CO_2-free air because carbon dioxide intake causes more rapid and deeper breathing. Based on rodent experiments, Levin et al. [16] found a 50 percent increase in carbon monoxide uptake when 5 percent carbon dioxide was present.

Hydrogen Cyanide

Hydrogen cyanide gas (HCN), which is 10 to 40 times as toxic as carbon monoxide [10], depending on concentration levels, is found in smoke from combustibles containing organically bound nitrogen, especially when burned with limited ventilation. Examples of such combustibles are wool, nylon, acrylonitrile polymers, urea-formaldehyde polymers, and polyurethanes.

It has been reported [7] that monkeys become incapacitated in 19 minutes by as little as 100 ppm of HCN in air. In other tests with mice, 177 ppm proved fatal in 29 minutes. [9]

The mode of action of hydrogen cyanide has been established. The hemoglobin in the blood picks up oxygen in the lungs and transports it throughout the body, where it is "unloaded" to the body cells and, most critically, to the brain cells. The process by which the cells accept the oxygen requires enzymes to catalyze it. If hydrogen cyanide is present, it does not interfere with the hemoglobin transport of the oxygen. However, after entering the bloodstream through the lungs, the hydrogen cyanide combines with the enzymes in the cells and inactivates them, so the cells can no longer accept oxygen. As little as 0.2 ppm by weight of HCN in the blood will cause toxic symptoms (salivation, nausea, confusion, and giddiness) and greater amounts will cause convulsion, coma, and death. For mice, 1 ppm by weight in the blood is lethal. [9]

Hartzell [14] proposed that a concentration-time product of about 1500 ppm-minutes in the air supply is likely to be hazardous to humans. This proposal implies that Haber's rule is valid for hydrogen cyanide. However, Purser [10] notes that hydrogen cyanide does not follow Haber's rule nearly as well as carbon monoxide does.

Hydrogen Chloride

Both carbon monoxide and hydrogen cyanide are narcotic (or asphyxiant) gases, and both act, in different ways, to interfere with oxygen uptake. However, other gases, such as hydrogen chloride, are powerful irritants, and have entirely different effects.

Polyvinyl chloride, as well as other organic compounds containing chlorine and hydrogen, will decompose in a fire in such a manner that most of the chlorine ends up in the smoke as hydrogen chloride. This method of decomposition can be contrasted with carbon monoxide and hydrogen cyanide, which, under some conditions, are largely destroyed by oxidation in the fire.

As little as 75 ppm of HCl is extremely irritating to the eyes and upper respiratory tract. Concentrations near 1000 ppm are very rapidly fatal to rats. [7] However, baboons exposed to concentrations well above 1000 ppm of HCl in air were not incapacitated immediately, but post-exposure deaths were caused due to lung damage. [14] Neither the potentially incapacitating nor lethal effects of hydrogen chloride on humans are as well understood as those of carbon monoxide and hydrogen cyanide. However Purser [10] has suggested that an exposure to approximately 1000 ppm of HCl for 20 minutes can cause permanent lung damage to humans or endanger life.

If a person breathes air containing a low concentration of hydrogen chloride, it will be absorbed mostly in the upper respiratory tract, and very little will reach the lungs. However, if smoke particles are present, much of the hydrogen chloride is likely to be adsorbed on the particles and could therefore be carried into the lungs with the particles. It is also possible that smoke containing hydrogen chloride could, on cooling, form an acid mist (aqueous HCl solution), which could settle out of the smoke or become adsorbed by nearby surfaces before being breathed. These complexities compound the problem of predicting how serious is the threat of hydrogen chloride in smoke.

Other Toxic Species

Acrolein (CH_2=CH—CHO) is an extremely irritant vapor produced in pyrolysis of various organic compounds, especially wood, paper, and cotton. Concentrations as low as 1 ppm cause marked irritation of the eyes and nose. Purser [10] proposed that the lethal concentration for humans can be between 80 and 260 ppm, for a 20-minute exposure. A concentration of about 5 ppm would produce a severe, but not necessarily unbearable, irritant effect.

Many other gases occasionally found in smoke are believed to be harmful. These gases, which generally are not found in lethal concentrations except when special combustibles are involved in the fire, include oxides of nitrogen (NO, NO_2), ammonia (NH_3), sulfur dioxide (SO_2), hydrogen sulfide (H_2S), hydrogen fluoride, hydrogen bromide, isocyanates, formaldehyde, and phosphorus compounds. Refer to Purser [10] for data on the effects of these and other gases.

Oxygen Deficiency

Even if nothing but carbon dioxide, water vapor, and nitrogen were present in the fire products, and these were to mix with the air being breathed by a fire victim, then the oxygen percentage would be reduced below the normal 21 percent. Very little difficulty is caused by short-term exposure to oxygen/nitrogen mixtures down to about 15 percent oxygen. Below this point, however, symptoms of lethargy, poor coordination, and confused thinking appear. In the range of 7 to 10 percent oxygen, unconsciousness and ultimately death occur.

Inasmuch as two of the principal toxicants in smoke—carbon monoxide and hydrogen cyanide—act to deprive the brain of oxygen, their effects would be enhanced if the oxygen level in the air were significantly below 21 percent. [9] The necessity of administering oxygen promptly to persons rescued from fires is obvious. *(See Figure 11.5.)*

Mixtures of Gases and Particulates

Smoke invariably will contain a deficit of oxygen, and some concentrations of carbon monoxide and carbon dioxide gases. In certain cases, hydrogen cya-

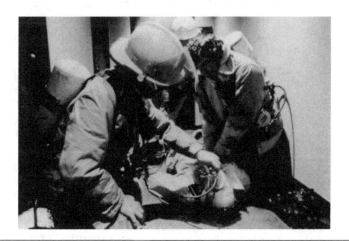

FIGURE 11.5 Fire fighter administering oxygen to fire victim.

nide also will be present. The combined effect of CO, CO_2, HCN, and low O_2 is to reduce the oxygen uptake of the body and, most critically, the oxygen uptake of brain cells. With a limited degree of success, mathematical formulas have been proposed that combine the individual effects of these four gases to predict the effect of various lengths of exposure to smoke on humans or test animals. [9, 10, 15, 17]

The starting point for these mathematical formulas is the application of Haber's rule. For a given length of exposure, if a mixture contains a fraction, x, of the lethal (or incapacitating) dose of carbon monoxide in air and a fraction, y, of the lethal (or incapacitating) dose of hydrogen cyanide in air, then the combination will be lethal (or incapacitating) if $x + y$ is greater than 1. Corrections are then applied for the effect of the increased breathing rate caused by carbon dioxide and the enhancement of the narcotic effect caused by the less-than-normal oxygen level.

Such a formula can be tested by exposing rodents to smoke from halogen-free combustibles and measuring the CO, CO_2, HCN, and O_2 content of the smoke. Such tests generally give an approximate, but not very close, agreement with the prediction. Some of the discrepancies can be due to other toxicants present. Other discrepancies probably are caused by the fact that, for short exposures at high concentrations, hydrogen cyanide is about 40 times as toxic as carbon monoxide, while for longer exposures at lower concentrations, hydrogen cyanide is only about 10 times as toxic as carbon monoxide.

When halogens are present in the combustible, giving rise to HF, HCl, or HBr in the smoke, then the problem exists of devising a single formula that combines the irritant effect and subsequent lung damage caused by these gases with the narcotic effect of CO and HCN. There is also the problem of dealing with other irritants, such as acrolein, formaldehyde, and oxides of nitrogen. The ability of smoke particles to absorb and act as a carrier for irritants requires more study. Again, while carbon dioxide accelerates breathing rate, irritants tend to reduce breathing rate, and the treatment of these opposing effects in a mathematical formula poses a problem.

In summary, carbon monoxide and sometimes hydrogen cyanide are believed to be the chief toxicants in most cases of fire. The effects of CO and HCN can be predicted approximately by formula when their concentrations and exposure times are known, with corrections for carbon dioxide and low-oxygen contributions. However, not enough is known quantitatively about the added effects of irritant gases, which can be quite significant in some cases.

Toxicity of Smoke from Specific Materials

Approximately six test procedures exist in various countries for burning small samples of pure or composite materials under controlled conditions and exposing a group of rodents to the smoke. [10] The test is repeated several times with different sizes of sample, to determine the weight of sample that will kill half the rodents, either immediately or within a specified period after the exposure. Typically, the critical sample size is found to be between 5 and 50 mg/L of the air present.

The data are not presented here because the results obtained from different test procedures do not agree with one another. The reasons are not understood fully. Clarke [18] reported that one test procedure showed red oak to be 3.4 times as toxic as polystyrene (based on quantity needed to kill one-half the rodents), while polystyrene was deemed 1.7 times as toxic as red oak by another test procedure.

In all these test procedures, the samples are heated to induce pyrolysis and burning. The procedures differ in the following ways:

1. The rate at which the samples are heated
2. The distance between the sample and the test animals
3. The quantity of air supplied
4. Whether the smoke accumulates in a closed space containing the animals or whether it flows continuously past the animals
5. The duration of exposure
6. How long the animals are observed after the exposure
7. Whether rats or mice are used
8. The temperature near the animals

Unfortunately, these differences seem to affect the results significantly. The only correct way to perform the test would be to duplicate the conditions existing in an actual fire. However, actual fires differ widely from one another. In particular, fires with restricted ventilation will release smoke containing more toxic products, especially carbon monoxide. Furthermore, smoldering or pyrolysis clearly give different products than flaming combustion of the same material. In view of these considerations, certain limited conclusions can be made about the various animal tests (bio-assays) performed to date:

1. The majority of materials tested, both natural and synthetic products, can be burned to produce types of smoke that have toxicity ratings differing from one another by a factor of about 10. Within this group, quantitative comparisons of one material with another are not very meaningful because of problems with variability of fires and with bio-assay procedures discussed above.
2. Nitrogen-containing polymers that produce substantial concentrations of hydrogen cyanide in the smoke are significantly more toxic in the bio-assays than nitrogen-free materials.

3. Insofar as effects of HCl or other halogen acids are concerned, rodents are believed to be poor surrogates for humans because of their different olfactory equipment.

4. Polytetrafluoroethylene (Teflon) normally will not burn in air, but if it is pyrolyzed at a high temperature in air, the products have far higher toxicity to rodents than do other substances. It is not certain what the toxic agent is. Teflon is the only known example of a "supertoxicant" (toxicity hundreds of times that of wood) among commercially used materials tested. There is no clear evidence of human fatalities caused by Teflon fumes.

5. Bio-assay results have been successfully correlated, at least approximately, with the known toxic effects of CO, CO_2, HCN, and low O_2, for a wide variety of materials. The situation for halogen-containing materials is less clear.

Dilution of Toxic Smoke

If a fire were burning in the open, air would continue to mix into the fire gases as they rose by a process called turbulent entrainment. The gases on the centerline of the plume ultimately would become diluted to a point where the concentrations of toxicants would be tolerable for a given period of exposure. Figure 11.6 shows how much the gases would have to be diluted to reduce the carbon monoxide concentration below 2400 ppm, assuming that the CO/CO_2 molar ratio (or volume ratio) generated by combustion is known. This ratio is influenced strongly by the degree of ventilation of the fire.

The dilution ratio in Figure 11.6 is simply the ratio of the volume of mixed-in air to the calculated volume of stoichiometric air, for a given quantity of burning material. It is assumed that the smoke has cooled to a breathable temperature and the water vapor generated by combustion has condensed. The combustible is taken to have the formula $C_aH_bO_c$, and the calculated curve depends on the ratio:

$$\frac{b - 2c}{a}$$

If the combustion is such that the CO/CO_2 ratio is very small or zero, it is still necessary to dilute the products sufficiently to provide at least a minimum breathable oxygen level. For the calculation, it was assumed that 12 percent oxygen was required. Figure 11.6 shows that a dilution of at least 2.3 times stoichiometric is always needed because of the oxygen requirement. If the CO/CO_2 ratio is as large as 0.5, then 19 to 29 times the stoichiometric air is needed, depending on

$$\frac{b - 2c}{a}$$

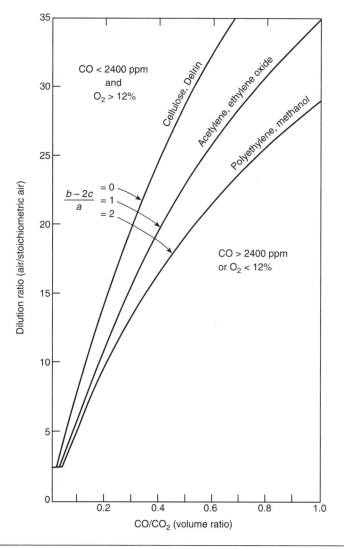

FIGURE 11.6 Dilution ratio required for combustion products of $C_aH_bO_c$ to achieve a given degree of breathability.

to keep the carbon monoxide level below 2400 ppm. Figure 11.6 is not meaningful if major proportions of toxicants other than carbon monoxide are present, unless they are represented as "CO equivalents."

Of course, concern with the toxicity of smoke relates to fires in compartments rather than fires in the open. When a fire is burning vigorously in a compartment, the upper portion of the compartment will contain a smoke layer with relatively little oxygen, so entrainment of air can occur only in the region from the bottom of the flame to the bottom of the smoke layer. This entrainment can be calculated if the size and intensity of the fire and the depth of the smoke layer are known. [3, 19] The depth of the smoke layer can be calculated from a knowledge of the fire intensity, the ventilation conditions in the compartment, and the rate of heat loss into the ceiling and walls of the compartment. [3, 20] Such calculations often require an elaborate computer program. *(See Chapter 13, "Computer Modeling of Fire.")*

In principle, the additional mixing with air can be calculated when the smoke emerges under the top of the doorway into a hall, and, in principle, additional calculations can be made as the smoke progresses down the hall and into other rooms, stairwells, and so on. Thus, it is theoretically possible to estimate the dilution occurring at each location, but it is difficult to perform such calculations accurately.

If an automatic sprinkler system began operating in the hall, it would be very effective in mixing and diluting the smoke, although it would reduce the visibility in the lower part of the hall.

NONTHERMAL FIRE DAMAGE TO MATERIALS AND EQUIPMENT

Fire not only kills or injures people, but also destroys or damages property, and can prevent the functioning of critical equipment. One mode of destruction, which will not be discussed here, is by heat or by collapse of the structure. Another means of damage is caused by a relatively small, localized fire in a large building, where smoke spreads to distant points in the building. The effects of this kind of damage from fire are often chemical in nature.

Along with damage by smoke, damage by extinguishing agents also should be considered. For example, dry powder agents contain salts that frequently are quite corrosive. Halogenated extinguishing agents, such as Halon 1301 (CF_3Br), while not corrosive themselves, will decompose in a fire to form HF and HBr, which are extremely corrosive. Also, water can cause damage in some cases (e.g., to books or documents), although vacuum-drying procedures have been developed to salvage most items.

Smoke moving through a building will be absorbed by any textiles encountered, and, as is well known, will impart an odor to these materials. Sometimes these odors can be removed by washing or dry cleaning. The chemistry associated with residual odors, however, is not well understood.

A fire involving polyvinyl chloride can create damage that is not immediately apparent. If such a fire burns in a reinforced concrete building, the HCl produced can possibly penetrate the somewhat porous concrete and attack the steel reinforcing rods. Corrosion can occur slowly, leading to structural failure at a later time.

If a fire involves an electric transformer or capacitor bank containing polychlorinated biphenyls (PCBs) inside a building, often the PCBs or their decomposition products will be transported and deposited in various places throughout the building. The structure of a PCB with 5 chlorine atoms per molecule is shown below. The number and locations of chlorine in the molecule can vary.

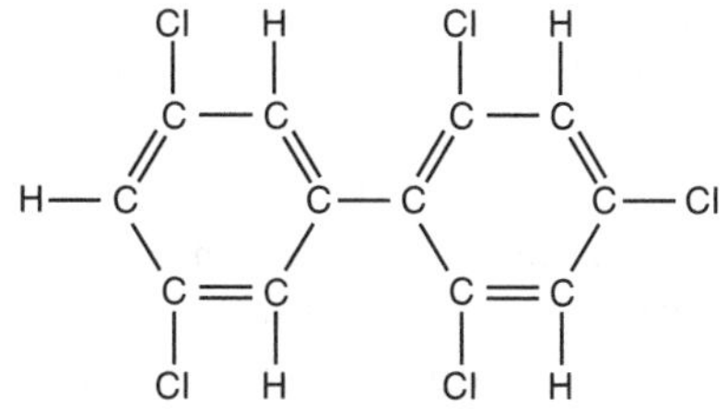

PCBs were found to be toxic, and their manufacture was banned in 1979. However, many previously installed units are still in use. Of even greater concern than the toxicity of PCBs is their possible transformation, in part, to the much more toxic chlorinated dioxin class of chemicals, an example of which is shown below,

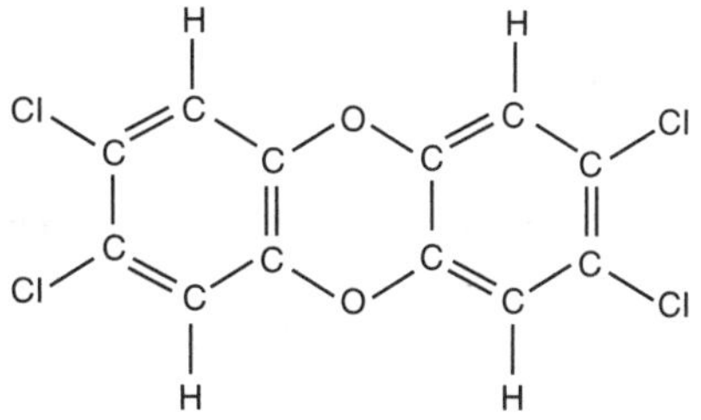

After a fire involving PCBs, a very expensive decontamination process of the entire building usually is required.

Perhaps the most important category of smoke damage is to electronic equipment, such as telephone exchanges, computers, and manufacturing plant control rooms. The most obvious type of damage is corrosion, i.e., the slow oxidation of metal exposed to air, which can be triggered by a substance in the smoke, often an acid. For example, if present in smoke, hydrogen chloride will attack most metals to form the metal chloride, which will promote further attack of the metal (catalysis). The presence of moisture or high humidity (greater than 40 percent) generally is necessary for rapid corrosion to occur.

A careful examination of the surface of a metal exposed to air under normal (nonfire) conditions often will show a chloride deposit up to 10 mg/m^2.

This amount usually is not harmful. [21] However, after exposure to smoke from a fire involving polyvinyl chloride, surface contamination of up to thousands of milligrams per square meter often is found, which often leads to significant damage.

Electronic equipment also can be rendered nonoperative because of a short circuit caused by conductive particles from smoke (carbonaceous matter), which can bridge a gap between conductors in a circuit.

Another possible effect of smoke is on electric contacts as in connectors or relays, where protective plastic coatings are not feasible. The introduction of foreign substances can cause the contacts of relays to stick.

Procedures for removing smoke contamination from electronic equipment have been developed; they include detergents, solvents, neutralizing agents, ultrasonic vibrations, and clean air jets. However, these procedures are largely empirical rather than scientifically based, and are not always effective. They sometimes give a temporary but not a permanent cure.

Problems

1. Smoke is sometimes black and sometimes white in appearance. Explain the reason for this difference.

2. Why is an exit sign in a smoke-filled room less visible when the ceiling light is more intense?

3. Of what use in fire protection is an infrared video camera?

4. What is the relationship between the particle size of smoke and the sensitivity of a smoke detector?

5. What type of smoke detector is more sensitive for a smoldering wood or cotton fire?

6. What are the most important toxic gases in smoke? Which are narcotic gases and which are irritant gases?

7. What is Haber's rule?

8. Explain why administering oxygen to fire survivors is so beneficial.

9. Why is it so difficult to interpret animal tests to determine accurately the toxicity of smoke from a given material?

10. In a closed room containing 40 m^3 of air and an electric fan to promote uniform mixing, how many kilograms of a typical combustible would have to burn to create an atmosphere that might cause a fatality in 30 minutes? Assume worst-case conditions.

11. What is a PCB? Why would it be present in a building?

12. Even though polyvinyl chloride and Halon 1301 are not themselves corrosive, explain how corrosion of metals can result if either material is involved in a fire.

References

1. Tewarson, A., "Generation of Heat and Chemical Compounds in Fires," *The SFPE Handbook of Fire Protection Engineering*, 1st ed., National Fire Protection Association, Quincy, MA, 1988, Section 1, pp. 179–199.

2. Mulholland, G. W., "Smoke Production and Properties," *The SFPE Handbook of Fire Protection Engineering*, 2nd ed., National Fire Protection Association, Quincy, MA, 1995, Section 2, pp. 217–227.

3. Drysdale, D., *An Introduction to Fire Dynamics*, 2nd ed., J. Wiley, New York, 1998.

4. Mulholland, G. W., V. Henzel, and V. Babrauskas, "The Effect of Scale on Smoke Emission," *Fire Safety Science: Proceedings of the Second International Symposium*. Hemisphere Publishing Company, New York, 1989, pp. 347–357.

5. Babrauskas, V., R. H. Harris, Jr., R. G. Gann, B. C. Levin, B. T. Lee, R. D. Peacock, M. Paabo, W. Twilley, M. F. Yoklavich, and H. M. Clark, "Fire Hazard Comparison of Fire-Retarded and Nonfire-Retarded Products," Special Publication 749, National Institute of Standards and Technology, Gaithersburg, MD, July 1988, p. 19.

6. Tewarson, A., and J. S. Newman, "Scale Effects on Fire Properties of Materials," *Fire Safety Science: Proceedings of the First International Symposium*, Hemisphere Publishing Company, New York, 1986, pp. 451–462.

7. *Fire and Smoke: Understanding the Hazards*, National Research Council-National Academy of Sciences, Washington, DC, 1986.

8. Birky, M., B. M. Halpin, Y. H. Caplan, R. S. Fisher, J. M. McAllister, and A. M. Dixon, "Fire Fatality Study," *Fire and Materials*, Vol. 3, 1979, pp. 211–217.

9. Esposito, F. M., and Y. Alarie, "Inhalation Toxicology of Carbon Monoxide and Hydrogen Cyanide Gases Released During the Thermal Decomposition of Polymers," *Journal of Fire Sciences*, Vol. 6, 1988, pp. 195–242.

10. Purser, D. A., "Toxicity Assessment of Combustion Products," *The SFPE Handbook of Fire Protection Engineering*, 2nd ed., National Fire Protection Association, Quincy, MA, 1995, Section 2, pp. 85–146.

11. Gottuk, D.T., R. J. Roby, and C. L. Beyler, "The Role of Temperature of Carbon Monoxide Production in Compartment Fires," *Fire Safety Journal*, Vol. 24, 1995, pp. 315–331.

12. Pitts, W. M., "The Global Equivalence Ratio Concept and the Formation Mechanisms of Carbon Monoxide in Enclosure Fires," *Progress in Energy and Combustion Science*, Vol. 21, 1995, pp. 197–237.

13. Pitts, W. M., "An Algorithm for Estimating Carbon Monoxide Formation in Enclosure Fires," *Proceedings of the Fifth International Symposium, International Association for Fire Safety Science* (available from Society of Fire Protection Engineers, 7315 Wisconsin Ave, Bethesda, MD 20814) 1997, pp. 535–546.

14. Hartzell, G. E., "Fire and Life Threat," *Fire and Materials,* Vol. 13, 1988, pp. 53–60.

15. Hartzell, G. E., D. N. Priest, and W. G. Switzer, "Mathematical Modeling of Toxicological Effects of Fire Gases," *Fire Safety Science: Proceedings of First International Symposium,* Hemisphere Publishing Company, New York, 1986, pp. 1059–1068.

16. Levin, B. C., M. Paabo, J. L. Gurman, S. E. Harris, and E. Braun, "Toxicological Interactions Between Carbon Monoxide and Carbon Dioxide," *Toxicology,* Vol. 47, 1987, pp. 135–164.

17. Levin, B. C., M. Paabo, J. L. Gurman, and S. E. Harris, "Effects of Exposure to Single and Multiple Combinations of the Predominant Toxic Gases and Low Oxygen Atmospheres Produced in Fires," *Fundamental and Applied Toxicology,* Vol. 9, 1987, pp. 236–250.

18. Clarke, F. B., "Toxicity of Combustion Products: Current Knowledge," *Fire Journal,* Sept. 1983, pp. 84–108.

19. Delichatsios, M. A., "Air Entrainment into Buoyant Jet Flames and Pool Fires," *The SFPE Handbook of Fire Protection Engineering,* National Fire Protection Association, Quincy, MA, 1988, Section 1, pp. 306–314.

20. Cooper, L. Y., "Compartment Fire-Generated Environment and Smoke Filling," *The SFPE Handbook of Fire Protection Engineering,* National Fire Protection Association, Quincy, MA, 1988, Section 2, pp. 116–138.

21. Sandmann, H., and G. Widmer, "The Corrosiveness of Fluoride-Containing Fire Gases on Selected Steel," *Fire and Materials,* Vol. 10, 1986, p. 4.

CHAPTER 12

Movement of Fire Gases

HEIGHTS OF DIFFUSION FLAMES

Imagine a combustible gas issuing upward from a pipe into air. If one were to move a probe gradually into the jet from the side, the probe would pass from pure air through a mixing zone consisting of progressively increasing combustible gas concentrations, until finally, if not too high above the top of the pipe, the probe would be immersed in pure combustible. A certain portion of this mixing zone surrounding the jet must contain mixtures between the lower and upper flammability limits.

If an ignition source is provided anywhere in the combustible zone, then ignition will occur, and the flame will spread to surround the combustible jet, at a speed consistent with the burning velocity. This process will be complete in no more than 1 or 2 seconds, unless the jet is extremely large. From then on, a diffusion flame will continue to burn indefinitely, as long as combustible gas is supplied from the pipe. The oxygen from the air will continue to diffuse through the combustion products to the flame.

Predicting the height of such a flame is possible. This flame height is a quantitative characteristic that is of practical importance in many fire situations. The flame height will depend on whether the flame is laminar or turbulent. Short flames generally will be laminar, tall flames turbulent.

Theories have been derived and confirmed that show that laminar diffusion flame heights are directly proportional to the volumetric flow rate of the combustible and are nearly independent of the pipe diameter and the nature of the combustible gas. (Hydrogen is an exception to this rule because of its high diffusion rate, which causes the flame to be shorter.)

To illustrate some magnitudes, a laminar methane diffusion flame about 9 cm (3.5 in.) high will be obtained if methane issues from a tube at the rate of 6 cm^3/s (0.8 ft^3/hr), regardless of the diameter of the tube. Lower or higher volumetric flow rates will give proportionately shorter or taller flames. The situation would be about the same for any other combustible gas (except hydrogen), because the flame height is controlled by the rate of diffusional mixing with oxygen rather than by chemical reaction rates.

A hazardous fire generally will be much larger than these laminar flames, and it will be turbulent. *(See Figure 12.1.)* The heights of turbulent diffusion flames are governed by different principles. There are two types of turbulent diffusion flames:

1. Momentum-dominated
2. Buoyancy-dominated

The former is associated with a high velocity jet issuing from an opening; the latter is exemplified by the "lazy" flame over a burning dish of flammable liquid, where the upward flow is driven entirely by buoyancy.

The flame height of a *momentum-dominated* jet is proportional to the diameter of the orifice and is independent of the flow rate, as long as the flow rate is great enough to maintain turbulence. This remarkable independence of flame height on flow rate is explained by the argument that the turbulence increases as the velocity is increased, promoting mixing with air at a proportionally greater rate. The flame height of a momentum-dominated diffusion

FIGURE 12.1 Burning tank car, showing buoyancy-dominated fire plume.

flame is from 50 to 300 times the burner diameter, depending mainly on the stoichiometric air–fuel ratio of the combustible.

The height, h, of a *buoyancy-dominated* turbulent diffusion flame, more often encountered in a fire, is correlated most successfully in terms of the convective heat-release rate, Q_c, of the flame, which is expressed in kilowatts. The formula

$$h = 0.23(Q_c)^{2/5} - 1.02d \qquad (12.1)$$

was proposed by Heskestad [1] and works for a wide variety of fire sizes and combustibles, including hydrogen. In this formula, h is the fire height in meters, and d is the diameter of the base of the fire in meters.

Solutions to this equation are plotted in Figure 12.2. The ratio h/d can vary at least from 1 to 44. When conditions are such that h/d is less than 1, the flame breaks up into a number of small flamelets. Of course, the formula cannot be used until the value of the convective heat-release rate is known for a particular fire. (See Chapter 10, "Fire Characteristics: Solid Combustibles." Also see Table 10.5.)

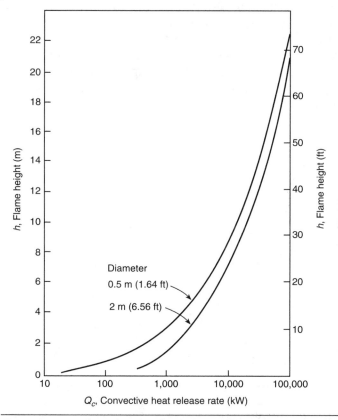

FIGURE 12.2 Flame height of turbulent diffusion flames versus the convective heat release rate (for 0.5-m and 2-m fire base diameters).

DETAILS OF THE STRUCTURE OF THE FIRE PLUME

Fire Plume in the Open

This subsection addresses a *fire plume* in the open. A fire plume is a buoyantly rising column of combustion products, along with not-yet-burned fuel vapor and admixed air. For a fire in a building, the plume will impinge on the ceiling, except for a very small fire or a very high ceiling. The intersection of a plume with a ceiling is discussed in the next subsection.

Figure 12.3 shows a turbulent column of hot gases rising because of buoyancy. The effect of the turbulence is to cause rapid mixing of the hot gases with the surrounding cooler air. The addition of cold mass to the rising column causes its velocity to decrease, as well as to make the column wider, and to reduce the temperature. When plume height is large in comparison to the width of the base of the plume, the average midline temperature (relative to the ambient temperature) is found to decrease at a rate inversely proportional to the 5/3 power of the height. The average midline velocity decreases more slowly, at a rate inversely proportional to the $1/3$ power of the height. The thickness of the plume increases at a rate directly proportional to the height. Formulas have been developed for predicting the temperature and velocity distributions across a plume at any given height, in terms of the heat release rate driving the plume.

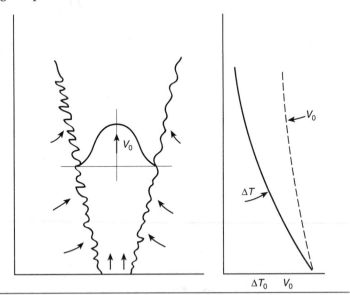

FIGURE 12.3   A buoyant turbulent plume, showing air entrainment velocity profile, and decrease of centerline temperature and velocity with height.

The foregoing relations refer to a rising column of hot gas, with no combustion taking place. This is applicable to a fire in which the combustion occurs close to the base of the fire plume. However, if combustion continues in the fire plume, the release of heat will cause an increase in temperature and velocity.

The turbulence intensity in a fire plume is quite high; the velocity fluctuations at the centerline can be of the order of 30 percent of the average velocity. The temperature fluctuations can be even greater.

In general, a fire plume will contain smoke particles. As surrounding air mixes into the plume, it dilutes the smoke as well as reduces the temperature. This mixing is called *entrainment*. To predict which environment will be produced by a given fire, it is necessary to know the rate of entrainment into the plume. While it is true that formulas have been proposed to calculate the rate of entrainment in a fire plume, the results are not completely reliable, because small ambient disturbances in the air near the plume can have substantial effects on the entrainment rate.

A rough approximation of the entrainment rate m' (kilograms/second), for a turbulent plume of height z (meters) and surface area A (square meters), is given by

$$m' = 0.188Az^{1/2} \qquad (12.2)$$

In a plume where combustion occurs only in the lower part, it is generally found that, at the plume height above which no combustion occurs, there is roughly an order of magnitude more entrained air present than the stoichiometric requirement.

Fire Plume Under a Ceiling

It is common for a fire plume impinging on a ceiling to make a 90-degree turn and spread out radially under the ceiling, forming what is called a *ceiling jet*. This ceiling jet is important for at least two reasons:

1. Devices to detect the fire, as well as automatic sprinklers, are generally mounted just under the ceiling, and knowledge of the time of arrival and properties of the ceiling jet are crucial for predicting when a detection device will be actuated.

2. The downward thermal radiation from the ceiling jet, and, a little later, from the hot ceiling itself, is a major factor in preheating and igniting combustible items not yet involved in the fire. This radiation is very important in affecting the rate of fire spread.

Two situations are possible:

1. The fire axis may be close to a wall.

2. The fire axis may be far from the nearest wall.

Figure 12.4 shows a turbulent ceiling jet, with the walls remote. The velocity of the jet will progressively decrease in the radial direction, because the flow is spreading radially, and the principle of conservation of momentum requires that the velocity must decrease. Further, the ceiling jet is generally turbulent, and mixing occurs between the jet and the air below. This mixed-in air slows down the jet and also reduces its temperature. The temperature is also reduced because of heat transfer from the jet to the ceiling. (The foregoing statements are valid only for the case in which the combustion is complete before the plume reaches the ceiling.)

Formulas have been developed for calculating the temperature and velocity distribution in a ceiling jet. For example, consider a region of a ceiling jet at a radial distance from the fire axis equal to the vertical distance from the fire source to the ceiling. In this region, the maximum velocity in the jet will have dropped to half the value near the fire axis, and the temperature (relative to ambient) will have dropped to about 40 percent of the value near the fire axis. The maxima of velocity and temperature will exist at a distance below the ceiling equal to about 1 percent of the distance from the fire source to the ceiling.

If the walls are much farther away than this, the temperature and velocity of the ceiling jet will have decayed to negligibly low values before the jet encounters the nearest wall. However, if the nearest wall is not far away, a reflection will occur when the jet reaches the wall. The reflected jet moves

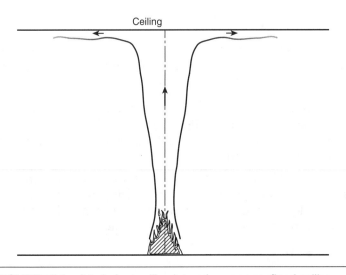

FIGURE 12.4   A turbulent ceiling jet under an unconfined ceiling.

back toward the fire axis, just under the original jet. Thus, the hot layer under the ceiling becomes thicker.

If the compartment has an open door or window and the fire continues, the hot layer will ultimately become thick enough to extend below the top of the opening, after which the hot, smoke-laden gases will start to exit from the compartment. The phenomena involved in this process are discussed in the next section.

If the fire is located next to a wall (or a corner), the behavior of the ceiling jet can still be predicted if it is assumed that the fire is twice (or four times) the actual size.

THE FILLING OF A FIRE COMPARTMENT BY SMOKE

If a fire is ignited in a compartment without openings, one of two things will happen.

1. The release of heat will cause an increase in pressure of the gases in the compartment, according to the *perfect gas law,* and this increased pressure may cause a rupture of the confining surfaces.

2. If no rupture occurs, the oxygen in the compartment will become exhausted, and the combustion will cease.

A more common case is that of a compartment with an opening. Figure 12.5 shows a fire in a compartment with an open doorway. The smoke has descended to a level somewhat below the top of the doorway, and at this point a steady state has been reached. (The assumption is made here that the fire is burning steadily but not growing.)

Knowledge about the hot layer is crucial to assessment of life safety in a fire compartment. Specifically, its thickness and temperature must be known,

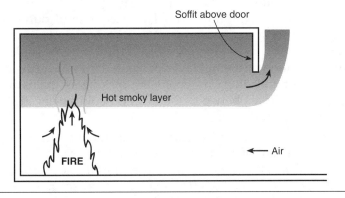

FIGURE 12.5 A steadily burning fire in a compartment with an open doorway.

both of which affect the intensity of downward radiation impinging on the lower part of the compartment. The rate of outflow of the hot layer must be known to assess fire safety in the adjacent space. First, the steady-state situation depicted in the figure must be quantified. Then, the time required to reach this condition must be studied.

It is obvious that the answers to these questions are crucial to assessment of life safety in a fire compartment. Enough is now known about the physics of a fire to permit at least approximate calculations giving quantitative answers to these questions.

First, consider the physics involved for the steady-state situation depicted in Figure 12.5. Air is entering the compartment through the lower part of the doorway. Air is being entrained into the fire plume and is being drawn upward into the hot gas layer. Hot gas is escaping from the compartment through the upper part of the doorway. These inward and outward flows are driven by pressure differences. How do these pressure differences come about?

The difference in pressure between the top and bottom of a column of gas of height h is equal to $g\rho h$, where ρ is the gas density. (See Chapter 4, "Flow of Fluids.") The gas in the hot layer has a substantially lower density than the air in the lower part of the room or the air outside. The resulting pressure variations with height are shown in Figure 12.6. In the doorway, there is a neutral height where the inside and outside pressures are the same. Above this height, the pressure is higher inside, causing an outflow. Below the neutral height, the reverse is true, and there is an inflow. The pressure variations can be calculated. The inflow and outflow rates can then be calculated, being proportional to the area available and the square root of the pressure difference.

The other factor is the rate at which air is being entrained into the fire plume. This rate is proportional to the size of the fire and the vertical distance from the base of the plume to the hot layer. The greater the entrainment rate, the lower the bottom of the hot layer. The mass rate of outflow is slightly greater than the rate of air inflow because of the mass contributed by the vaporization or pyrolysis of combustible. Thus, it is possible to make a steady-state calculation of the hot layer position for a given fire intensity, and given compartment and doorway geometry, as long as the assumption of a sharp horizontal interface between the hot and cold layers in the compartment is more or less valid.

It is more difficult, but also possible, to calculate the transient situation, starting from the instant of ignition, if the fire intensity and its variation with time are known. When the fire first starts, the air entrained into the fire plume expands and rises to the ceiling. The expansion causes a pressure increase throughout the chamber, which causes air to flow out the doorway. The outward flow occurs both near the bottom and near the top of the doorway.

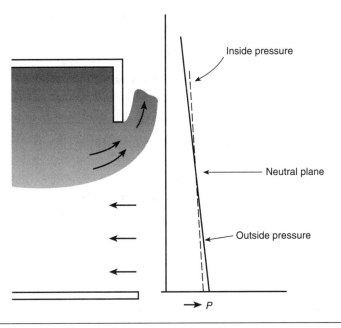

FIGURE 12.6 Pressure gradients at the doorway *(see Figure 12.5)*, caused by relative densities of cold air and hot gases.

The ceiling jet reaches the walls, and then the hot layer progressively becomes deeper, the rate of descent depending on the fire entrainment rate and on the size of the compartment. When the hot layer becomes deeper than the top of the doorway, hot gases start spilling out. Air starts flowing in, as shown in Figure 12.5. Finally, if the fire does not continue to grow, a steady state is reached, as discussed previously. Even if the fire does continue to spread, a "choked" condition will be reached at which the combustion is limited by the rate at which air can enter. This corresponds to the hot layer extending nearly to the floor.

Calculations of this transient process can be made, but they are sufficiently complex that they can best be done with a high-speed computer. Such computer programs, based on so-called *zone models*, are discussed in Chapter 13, "Computer Modeling of Fire." However, if the conditions in the fire compartment are such that the assumption of a horizontal plane separating uniform hot and cold layers is not realistic, then a more elaborate field model must be used.

The choked condition described earlier can also be referred to as a post-flashover condition. For this case, the flow of air into the compartment and hot gases out can be described very simply, albeit approximately, in terms of the size of the door or window opening. If the opening has height H (meters) and area A (square meters), then the mass rate of flow in or out m' (kg per second) is given by

$$m' = 0.09AH^{1/2} \qquad (12.3)$$

This equation has been validated for wood cribs burning in compartments with various ventilation openings, but may not be valid for other combustibles.

SMOKE MOVEMENT IN BUILDINGS

If a building consisted simply of a series of compartments connected by open doorways, all on the same level, it would be a simple matter to use the same methods just described for a single compartment, calculating how the outflow from one compartment fills the next compartment. Such computer models exist. However, in most buildings there are additional complications:

1. In the calculation, it is necessary to know which doors or windows are open.
2. A window initially closed may break in the course of the fire.
3. The building may contain long corridors, and the assumption of a corridor filling from the top down is not realistic.
4. A multistory building will contain stairwells and possibly atria, with major effects on smoke movement.
5. An industrial building or warehouse may have a sloping ceiling and/or a ceiling supported by beams, which would influence the ceiling jet.
6. The presence of a wind outside a building will influence air movement within the building, if there are any leaks.
7. The operation of a heating, ventilation, or air-conditioning system will have a profound effect on smoke movement in a building. Even if the system is shut down, hot gases may still move through the ducts, because, aside from buoyancy, the heat released by a fire will cause an expansion of gas.
8. In a tall building, there is a *stack effect*. Imagine a tall building the interior temperature of which is warmer than the outside temperature (winter) or cooler than the outside temperature (summer, with air-conditioning). The pressure difference between the air outside the top of the building relative to the air outside the bottom of the building is proportional to the outside air density, while the corresponding pressure difference inside the building is proportional to the inside air density. As a result, in winter, air would leak into the building at the lower levels and leak out at the upper levels, prior to the occurrence of any fire. The resulting upward flow inside the building would help carry smoke

upward, and, conversely, downward, in summer with air conditioning, once the smoke had cooled.

In view of all these complications, any program to calculate the movement of smoke through a building as the result of an assumed fire at an assumed location will be complex. For life safety reasons, it may be appropriate to make worst-case assumptions, which would tend to simplify the calculations.

The introductory material constituting this chapter may be supplemented by readings in much more detailed texts, such as *The SPFE Handbook of Fire Protection Engineering*. [2]

Problems

1. A fire with a circular base 1 m in diameter has an intensity of 600 kW, of which 70 percent is convective and 30 percent is radiative. Estimate the height of the flame.

2. Calculate the rate of air entrainment into the fire plume in Problem 1 from the base to the flame tip.

3. How does the presence of a ceiling affect a fire plume?

4. What is meant by choked flow at an opening of a fire compartment?

5. Why might the movement of smoke from a fire in a tall building be different in summer and winter?

References

1. Heskestad, G., "Luminous Heights of Turbulent Diffusion Flames," *Fire Safety Journal,* Vol. 5, 1983, pp. 103–108.

2. DiNenno, P.J., *The SFPE Handbook of Fire Protection Engineering,* 2nd ed., National Fire Protection Association, Quincy, MA, 1995.

CHAPTER 13

Computer Modeling of Fire

TYPES OF MODELS

Computer models have been developed to calculate a variety of aspects of fires. They can give results in rough agreement with experiment in certain cases, but not in other cases. In 1992 Friedman [1] listed 62 existing models. By 1998 the number of models being generated in a dozen countries must approach 100.

The largest class of models addresses a developing fire in a compartment or a series of interconnected compartments (a building). It is assumed that the model user knows the following properties in advance:

1. The compartment and building geometry
2. The ventilation conditions
3. The types and locations of the combustible materials present
4. The ignition site

For almost all models, the model user must be able to specify the burning rate of the first ignited object or at least how it would burn if in the open (i.e., well-ventilated; no radiative feedback from the surroundings).

Then the model calculates the distribution of temperatures, smoke, and toxic gases in the fire compartment and in connecting compartments, perhaps each second. A class of submodels of this type of model deals with the time required to actuate detectors at various locations.

Another class of submodels calculates the time needed to evacuate buildings. The user must know the initial locations of the occupants. The fire model, together with the detector submodel, gives the time available for escape, while the evacuation submodel gives the time needed for escape.

Another class of models deals with fire endurance. If a fire continues for a long time, say in the basement of a building, the heat may cause failure of steel or reinforced concrete structural elements. The model calculates how long an exposed structural element can survive before collapse.

Other models have been developed for a variety of fire problems. There are at least five models dealing with the interaction of water spray from sprinklers with a fire. The object is to predict when and if fire control is achieved.

Models have been developed to predict rate of flame spread, fire plumes extending from windows, smoke in ventilation systems, and fires in mine shafts. The following discussion concentrates on the largest class of model: the development of fire in a compartment.

USERS OF MODELS

In general, users of computer models can be classified in two groups. The first kind of user is concerned with better design procedures or guidelines for fire-safe structures or vehicles. The second kind is interested in reconstruction of details of actual fires.

The first type of user might ask the following questions: In a large building, what gain in life safety is obtained by providing an extra stairwell? Or providing an automatic sprinkler system? Or replacing the furniture or wall coverings with more fire-resistant materials? How many detectors should be installed and where? What should be the capacity of a fan for smoke removal? More generally, what modifications of an existing structure are needed, if any, to result in a high probability of very low casualties from a fire? Could a computer fire model be a key element in a building code?

The second type of model user is usually a litigant, or an arson investigator, or possibly a building code official. It is often the case that reliable observers are not available, especially during the early stages of a fire; a computer model can be very helpful in bridging the gap between the origin of the fire and what was observed later or could be deduced from a postfire study of the remains. The benefits to litigators or arson investigators are obvious. In the case where a building that conformed to the building code nevertheless suffered a major fire, it is of great interest to code officials to determine what went wrong.

FIELD MODELS

Field models predict two-dimensional or three-dimensional distributions of velocity and temperature. Figure 13.1 is an example of a field model calculation of the flow induced in a compartment by a fire of assumed intensity. The length of each arrow represents the velocity at that point. The low-velocity inflow of air into the compartment and the higher velocity outflow of combustion products under the door soffit are seen. The computer also calculates the temperature and gas composition at each point. Such a calculation requires a very powerful computer.

Figure 13.2 is another example of a field model computation. A fire of an assumed growth rate occurs at the closed end of a shopping mall, and the combustion products advance along the mall, around a corner, and ultimately out of the open far end. The interface between hotter and cooler gas is shown. Vision is obscured above but not below this interface.

Bilger [4] reviewed the status of field modeling and predicted that one day, with more advanced computers and more efficient calculational methods, field modeling would be more widely used by practitioners of fire protection.

ZONE MODELS

Zone models cannot predict multidimensional distributions of velocity and temperature. They are based on the approximation that a fire compartment can be divided into an upper zone and a lower zone, separated by a horizontal plane. The temperature and composition are assumed to be uniform in each zone. Velocity differences at various locations in each zone are ignored, except that an assumed fire plume is considered to entrain air at a calculated rate, up to the plane separating the two zones. The average velocity of air entering or leaving the compartment and the average velocity of combustion products leaving the compartment are calculated by the model.

Since zone modeling is so much simpler to use than field modeling, requiring only a personal computer, it is widely used by fire protection engineers. The first step in using a zone model is to input the prefire conditions (e.g., room dimensions, ventilation, location of first ignited object). This task is straightforward. The second step, not so straightforward, is to input the fire itself into the model.

In general, the needed information must be obtained from fire tests performed under a calorimeter. (See Chapter 10, "Fire Characteristics: Solid Combustibles.") Often, fire test results for the specific item of interest will not be available, so an approximation of the fire behavior (heat release rate versus time) is made on the basis of available fire test results from somewhat similar items. Test results tabulated in *The SFPE Handbook of Fire Protection Engineering* [5] may be useful.

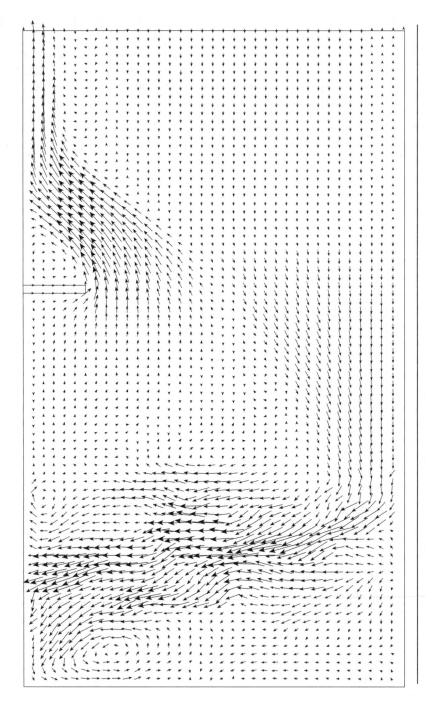

FIGURE 13.1 Field model calculation of fire-induced flow in a compartment with an open door. [2]

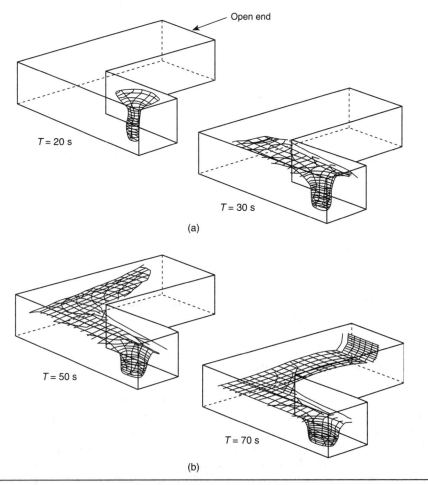

Open end

T = 20 s

T = 30 s

(a)

T = 50 s

T = 70 s

(b)

FIGURE 13.2   Temperature contours at various times (T) for a fire in a shopping mall. [3]

Figure 13.3 is an example of an assumed heat release rate. The assumed fire grows linearly with time for the first 100 seconds, reaching 800 kW, then continues steadily at this rate for another 200 seconds, after which burnout is assumed to occur. Suppose the burning object is an item of upholstered furniture. The heat release rate figures might represent an average of measurements made on a number of similar items, burning in the open.

An object burning in a fire compartment may burn faster than in the open because of radiant energy feedback from the hot smoke layer and the ceiling to the object. Or, it might burn more slowly than in the open, or stop burning altogether, because the "air" drawn into the base of the fire may have become

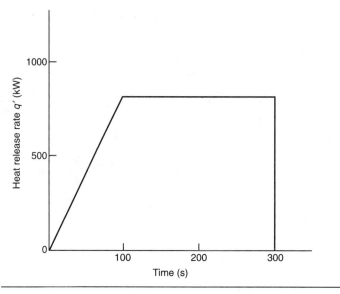

FIGURE 13.3 An assumed heat release rate for computer modeling.

diluted with combustion products. Some zone models attempt to take these effects into consideration.

After the assumed fire has been inputted into the model, the progress of the fire may be calculated, second by second if desired. A desktop computer may require only a few minutes. Some illustrative results are shown in Figure 13.4. The fire is burning on a furniture item in a compartment, with a soffit above the open door. There is an adjacent compartment, with the far end open. (In principle, the model could include any number of interconnected compartments, as well as forced ventilation ducts.)

As shown in Figure 13.4, at 20 seconds after ignition, the flame is still relatively small. A hot layer has begun to accumulate under the ceiling, but its temperature is only 100°C because the rising plume of hot gas from the fire is being diluted with cold air mixing (being entrained) into it. Also, the hot layer is losing heat to the cold ceiling. The computer is calculating these events. Note, however, that the computer predicts the hot layer to have the same temperature throughout, while it is obvious that the part of the hot layer directly over the fire will be hotter than in the far corner of the room.

Note that the air flow in the doorway at this time is all outward, because of the expansion of gases in the compartment caused by the heat released. At 100 seconds, the hot layer has become thicker, descending to a lower level, but is not yet below the top of the doorway. The temperature is now higher, mainly because the heat release rate is greater. All flow is still outward.

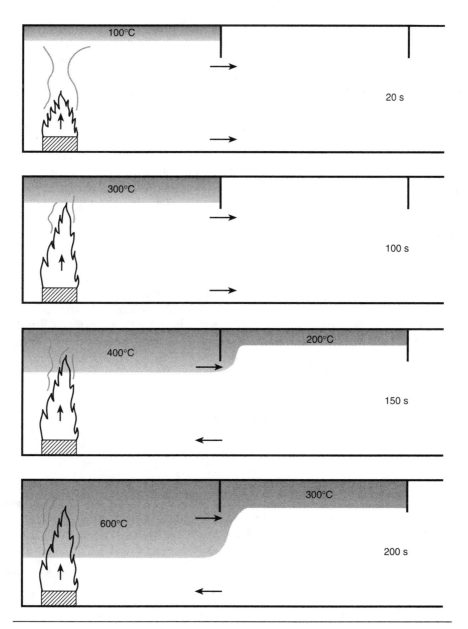

FIGURE 13.4 A representation of zone model calculations for fire in a compartment.

At 150 seconds, the hot layer is spilling out under the top of the doorway (the soffit) and is beginning to form a hot layer in the second compartment. The hot layer temperature in the fire compartment is now higher because some of the fire plume is submerged in the hot layer, so there is less entrainment of cool air into the plume. At the lower part of the doorway, air is now entering the fire compartment. The computer assumes that some mixing is occurring as the hot layer spills turbulently under the soffit, so the hot layer in the second compartment has been diluted and cooled by this mixing.

At 200 seconds, the hot layer has descended as low as it can. This condition will continue until burnout. The hot layer is accumulating in the second compartment and will soon start flowing into the third compartment.

At each stage of this process, the computer calculates the following:

1. Composition of the hot layer

2. Temperature

It is necessary for the model user to specify in advance what fraction of the combustible is going to be converted to soot or carbon monoxide, based on prior laboratory studies. This is easiest when the top of the flame is below the hot layer, but becomes more complicated when the flame extends into the hot layer. The computer can track the oxygen concentration as the plume enters the hot layer and can terminate combustion when a limit is reached, depending on temperature and oxygen concentration.

These outputs can determine the following:

1. Time a detector just outside the compartment will actuate

2. Visibility through the hot layer in the second compartment

3. Toxic potential

Another important effect must be mentioned. The soot particles in the hot layer will radiate heat. The ceiling, once it becomes hot, will also radiate heat. This heat may be enough to ignite a second flammable object in the fire compartment. The radiation can also prevent the survivability of people, even if they stay below the hot layer. Zone models can include such radiation calculations.

CFAST Zone Model

One of the most widely used zone models is *CFAST* (consolidated model of fire growth and smoke transport), developed at the Building and Fire Research Laboratory of the U.S. National Institute of Standards and Technology. [6] Some details of this model are presented.

CFAST can address a building with up to 30 compartments, all on one level, with multiple openings between the compartments and to the outside. Mechanical ventilation can be handled, with up to 30 ducts and 5 fans. The user must specify the fire heat release rate and the generation rates of the various combustion products. Multiple fires can be inputted. Heat transfer through walls and ceilings is calculated. Mixing at the plume and at the door or window is calculated.

The outputs include not only layer properties but also flow rates, radiative fluxes, and surface temperatures. The model stops combustion when oxygen becomes deficient.

CFAST does not treat stairwells or long corridors effectively. It does not include radiation feedback that causes increase in burning rate. As research continues, these features will presumably be added.

LIMITATIONS OF MODELS

The most obvious limitation of both zone and field models is the requirement for the user to supply the heat release rate of the fire. The necessary information is often hard to obtain.

Zone models are limited by their use of two horizontal layers without any temperature variation within a zone. Measurements in fire compartments often show departures from this behavior. Further, entrainment into the fire plume, mixing at the soffit, and any mixing between hot and cold layers must be handled by formulas incorporated into the zone model. Such formulas may not be accurate. (Field models can calculate these mixing processes without using such formulas, but approximations in the field models may lead to inaccuracies in the results.)

The following are the main limitations of the field models:

1. The need for very powerful computers
2. The need for highly qualified programmers to modify a field model for any application

Nonlinear behavior is another limitation in computer fire models. For example, a small error in the calculated temperature of the hot layer will cause a large error in the radiative heat flux impinging on a target because radiative emission varies with the fourth power of the absolute temperature.

Figure 13.5 illustrates another type of nonlinearity. Suppose that a radiative flux of 20 kW/m², from a hot layer, is impinging on a combustible target, which ignites spontaneously if it is heated to 430°C. The curve shows that this temperature will be reached in 105 seconds. Now, suppose there is a 10 percent error in the calculation of the radiative flux, which could be caused by a 3 percent error in the hot layer temperature. If the flux on the target is 18 instead of 20 kW/m² (10 percent less), then it is seen that the time to ignition is 225 seconds instead of 105 seconds. Thus, a small error in one quantity can cause a large change in a calculated quantity, in this case.

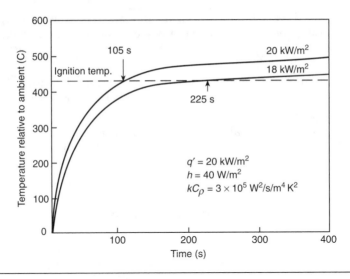

FIGURE 13.5 Temperature increase of a combustible slab subjected to radiative heating and convective cooling.

If the radiative flux had been 17 instead of 18 kW/m^2, ignition would not have occurred, even in infinite time. The target is being convectively cooled by the incoming cool air while it is being radiatively heated, so that, unless it ignites, it will reach an equilibrium temperature, which may be below the ignition temperature. For the example selected, the difference between 17 and 20 kW/m^2 is the difference between no ignition and ignition in 105 seconds. This would present a major challenge to any computer program trying to predict ignition of a second object.

Of course, if the radiative flux had been 40 kW/m^2 instead of 20 kW/m^2 (corresponding to a hot layer temperature of about 650°C instead of about 500°C, assuming an optically thick hot layer), this problem would disappear, and ignition would occur in about 24 seconds, plus or minus a few seconds because of errors in the flux. With a 40 kW/m^2 flux, the radiative flux would be much greater than the convective cooling at the ignition condition.

Other situations can lead to possible errors in computer calculations of fires. (See Friedman. [1]) Nevertheless, computer calculations offer a picture of fire development in a building that is far superior to what could be obtained by any other method. In many cases, computer predictions have been found to be reasonably close to actual fire behavior.

As fire science improves, it is reasonable to expect that computer predictions of fire will improve significantly.

Problems

1. What are the differences between computer zone models and computer field models?

2. For what reasons might a person want to use a computer fire model?

3. How might one obtain the needed burning rate input for a fire model?

4. State two reasons why the burning rate of an object in the open might be different from the burning rate of the same object in a compartment.

5. What type of fire model would be appropriate for studying smoke movement in a tall atrium of a hotel? In a stairwell?

6. How might a computer fire model be used as part of a building fire code?

References

1. Friedman, R., "An International Survey of Computer Models for Fire and Smoke," *J Fire Protection Engineering,* 4, 1992, pp. 81–92.

2. Walton, W. D., and E. K. Budnick,, "Deterministic Computer Fire Models," *Fire Protection Handbook,* 18th ed., National Fire Protection Association, Quincy, MA, 1997, pp. 11–53.

3. Stroup, D. W., "Using Field Modeling to Simulate Enclosure Fires," *The SPFE Handbook of Fire Protection Engineering,* 2nd ed., National Fire Protection Association, Quincy, MA, 1997, pp. 3–157.

4. Bilger, R. W., "Computational Field Models in Fire Research and Engineering," *Fire Safety Science—Proceedings of the Fourth International Symposium,* IAFSS, 1994. (Available from Society of Fire Protection Engineers, 7315 Wisconsin Ave., Bethesda, MD 20814.)

5. *The SFPE Handbook of Fire Protection Engineering,* 2nd ed., National Fire Protection Association, Quincy, MA, 1995.

6. Peacock, R. D., et al., "CFAST, the Consolidated Model of Fire Growth and Smoke Transport," U.S. Dept. of Commerce, Washington, D.C., NIST Tech. Note 1299, 1993.

CHAPTER 14

Fire-Fighting Procedures

EXTINGUISHING AGENTS

Water

Compared with any other fire-fighting agent, water is obviously low in cost and readily available. However, water also has some unusual properties that make it uniquely desirable as an extinguishing agent. An ideal liquid extinguishing agent should have the following characteristics:

1. It must be nonflammable itself.

2. It should have a high heat of vaporization. (The value for water, per unit mass, is at least four times as high as that of any other agent.)

3. Its boiling point should be well below the 250°C to 450°C range of pyrolysis temperatures for most solid combustibles. (Water boils at 100°C.)

4. It should be a mobile liquid at normal ambient temperatures, such as –20°C to +40°C. (Water is a mobile liquid except at below 0°C.)

5. It should be nontoxic and should not decompose to form toxic products. (Water is the only extinguishing agent that qualifies fully; liquid nitrogen or carbon dioxide can asphyxiate.)

6. It should cause minimal property damage. (Some, but not all, water-damaged items can be salvaged after a fire.)

7. It should not conduct electricity. (Water fails this test.)

Table 14.1 shows some properties of water and a variety of other nonflammable liquids that might be used as extinguishing agents (if cost considerations are ignored). Water is overwhelmingly superior to all other extinguishing liquids, primarily because of its high heat of vaporization, ideal boiling point (well above room temperature and well below the decomposition temperature of most combustibles), and nontoxicity. However, in some special situations, which will be discussed, other extinguishing agents might be required.

Water can be used as a solid stream from a fire hose; as a fog from a fog nozzle; as droplets from an automatic sprinkler *(see Figure 14.1)* or a spray nozzle; or as the major ingredient in fire-fighting foams, which are discussed later in this chapter. The hardware required for these uses is described in the *Fire Protection Handbook.* [1]

TABLE 14.1 Nonflammable liquids that could be considered as extinguishing agents

Liquid	Latent heat of vaporization (J/g)	Liquid range (m.p./b.p. in °C)	Comments
Water	2260	0/+100	
Carbon dioxide	573*	*	A, P
Mercury	292	−39/+357	T
1,2,4-Trichlorobenzene	263	+17/+213	T
Chloroform ($CHCl_3$)	247	−64/+61	T
Nitrogen	200	−210/−196	A, C
Bromine	194	−7/+59	T
Carbon tetrachloride	194	−23/+77	T
Argon	167	−189/−186	A, C
Halon 1301 (CF_3Br)	119	−168/−58	P
Perfluorobutane (C_4F_{10})	102	−128/+4	P

*Heat of sublimation at −78°C and 1 atm; carbon dioxide is a liquid only at elevated pressures (60 atm at +22°C, 5 atm at −57°C)
A = asphyxiant
C = cryogenic cooling required
P = pressurized storage required
T = toxic

FIGURE 14.1 An upright automatic sprinkler discharging water. Courtesy of Factory Mutual Research Corporation.

It is possible to improve the properties of water for certain applications by using additives. Addition of dissolved salts to water will lower the freezing point as well as enhance the ability to suppress certain fires. For example, addition of 466 g of calcium chloride (80 percent pure, with sodium dichromate corrosion inhibitor) to 1 L of water will lower the freezing point to –29°C (–20°F). Unfortunately, the inhibitor reduces but does not eliminate corrosion.

The freezing point of water can be lowered without introducing a corrosive substance by dissolving an organic compound, such as ethylene glycol $[C_2H_4(OH)_2]$ or glycerol $[C_3H_5(OH)_3]$ in water. A solution of 44 percent by volume of ethylene glycol and 56 percent water has a freezing point of –29°C (–20°F). If, after application to the fire, some of the water evaporates, the residual liquid will be more concentrated in ethylene glycol and could become flammable. Another problem with freezing-point additives is that they cannot be used in sprinkler piping that is connected directly to a public water supply. (In some locations, use is permissible if check valves are installed.)

Another type of additive is a *wetting agent*, or detergent, very small additions of which can lower the surface tension of water and improve its ability to penetrate porous materials, such as bales of cotton, stacked hay, or mattresses.

It is also possible to thicken water, that is, to increase its viscosity substantially, by use of additives. When water is thickened, it adheres more readily to a vertical surface of a combustible, rather than running off. Also, when projected from a nozzle, a stream of water containing a thickening additive remains longer as a coherent jet (before breaking up) and projects farther. *(See Figure 14.2.)*

However, there are problems with using thickened water:

1. Penetration of porous media is poorer.
2. Friction loss in a hose is greater.
3. Fine sprays cannot be obtained.
4. Floors become slippery.
5. The cleanup problem after a fire is magnified.

Accordingly, thickened water is not used frequently, except in fighting forest fires, where slippery floors and post fire cleanup are not issues. CMC (sodium carboxymethyl cellulose) and Gelgard® (a cross-linked polymer) are examples of commercial thickening agents.

Friction-reducing additives have different purposes. The diameter of a fire hose is limited by the strength of the fire fighter who must carry the hose up a stairwell. Reducing the friction of water flowing through a fire hose of a given diameter results in a higher water discharge rate for a given pumping pressure. Alternatively, a smaller diameter hose (or sprinkler piping system) could be used to give the same discharge rate. Polyethylene oxide is a water-

FIGURE 14.2 Coherent stream of water.

soluble linear polymer that, when added to water (1 part by volume of concentrate to 6000 parts of water), can give at least 40 percent greater water delivery from a hose.

If any kind of additive is to be used in an automatic sprinkler system or in a spray nozzle installation, a mechanical device usually is needed for proportionally mixing the additive with the flowing water. The reliability of such a proportioning device is never 100 percent, so the overall reliability of the protection system is reduced below that achievable if ordinary water is used. This consideration discourages the use of additives for fixed installations because they must be highly reliable. (If the water with premixed additive is to be taken from a storage tank, then this reliability problem disappears.)

The water discharged from an automatic sprinkler or a spray nozzle is in the form of droplets. The size distribution of these droplets is important. If the droplets are too small, they cannot penetrate to the seat of the fire but are carried upward by the fire plume. If they are too large, their surface-to-mass ratio is small and they cannot effectively cool the fire gases or the ceiling above the fire. The ideal drop-size distribution is thought to be a mixture of large and small drops with no intermediate-size drops (a bimodal distribution). Such a distribution is not easy to achieve. Figure 14.3 shows a representative drop-size distribution from a commercial automatic sprinkler. The mean drop size increases with the radial distance from the sprinkler axis.

The mean drop size of water from a sprinkler or spray nozzle depends on three variables:

1. The design and size of the device
2. The pressure drop, ΔP, across the device
3. The surface tension, σ, of the water

If D is a characteristic dimension of the spray device, then the mean drop size is approximately proportional to

$$\left(\frac{\sigma D^2}{\Delta P}\right)^{1/3}$$

The surface tension of water is 76 dynes/cm (0.076 N/m) at 0°C, 73 dynes/cm at 20°C, and 70 dynes/cm at 40°C. Addition of a dissolved salt would raise the surface tension slightly. For example, addition of 10 percent sodium chloride would raise the surface tension about 4 percent. However, addition of only 0.1 percent of sodium octyl sulfosuccinate, a wetting agent, would lower the surface tension from 73 to 28 dynes/cm. The lower the cube root of the surface tension, the smaller the drop size. This phenomenon has been used in research involving small-scale models of sprinklered fires.

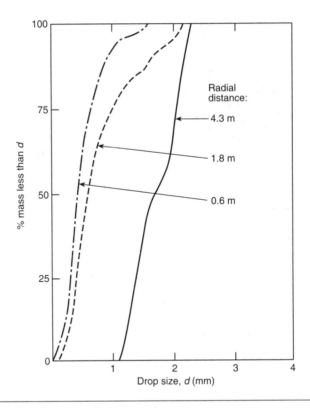

FIGURE 14.3 Drop-size distributions at various radial distances, 3 m below a commercial $^1/_2$-in. sprinkler operating at $\Delta P = 2$ atm. Source: You. [2]

The water-delivery rate required to control a fire can be expressed in units of volume of water delivered per unit floor area per unit time (m^3/m^2/min or gal/ft^2/min). Or, it can be expressed in millimeters or inches of water depth accumulated per minute, in the same way that rainfall is measured. One gal/ft^2/min is the same as 40.7 mm/min. The minimum rate needed depends on many variables, including the nature of the combustible. (See Extinguishment of Burning Solids in this chapter.)

Water Mist

Water mist extinguishing systems are currently being developed, largely because halogenated agents containing bromine, which have been widely used for the past 30 years, are no longer being produced, for environmental reasons. (See subsection entitled Halogenated Agents later in this chapter.) A system for applying a fine mist of water to a fire would be highly acceptable from an environmental viewpoint.

Three mechanisms exist by which a fine water mist might extinguish a fire:

1. The mist droplets, while evaporating, remove heat, either at the surface of the combustible, reducing pyrolysis, or within the flame, reducing the flame temperature.

2. The mist droplets evaporate in the hot environment, perhaps even before reaching the flame, generating steam, which acts as a diluent. Also, the expansion caused by evaporation may prevent air from entering a fire compartment.

3. The mist droplets block radiative heat transfer from the flame to the combustible.

In those tests in which mists have extinguished fires, some combination of these three mechanisms is believed to play a role. The situation is not fully understood at this time.

As to droplets cooling a burning surface, very fine drops (mist) would tend to be blown away from the surface by the pyrolysis gases. However, the drops could reach the surface if the mist is directed at the surface with high momentum.

In regard to cooling and diluting the flame, a sufficiently high concentration of water mist would have to be achieved. This is estimated to be around 15 percent water mist by weight in air. This could be achieved by spraying the mist directly at the flame. However, mist sprayed into the fire compartment but not directly into the flame would tend to settle to the floor, the settling rate depending on drop size. A drop of 10 µm diameter would settle 0.3 m (1 ft) in about 100 seconds, while a 50-µm drop would settle this distance in about 4 seconds. It is difficult to produce fine enough drops to build up an adequate concentration of mist in even a small compartment, with existing spray devices.

Blockage of radiation by a mist will often be effective in reducing the intensity or spread rate of a fire but will rarely be sufficient in itself to produce extinguishment.

In summary, the effectiveness of a fine mist will depend on the direction and momentum of the spray and on the compartment geometry. A number of documented fire tests have found mists to be effective. Testing is ongoing. The current status is reviewed by Mawhinney and Solomon. [3]

Aqueous Foams

The principal application of *aqueous* foam agents is for fighting flammable liquid fires. If the flammable liquid is lighter than water and is insoluble in water, then application of water would simply result in the liquid floating on water and continuing to burn. If the burning liquid is an oil or fat, the temperature of which is substantially above the boiling point of water, then the water will penetrate the hot oil, turn into steam below the surface, and cause an eruption of oil that will accelerate the burning rate and might spread the fire. Foams are the primary tools for fighting fires that involve substantial quantities of petroleum products, for example, refineries, tankers, and storage areas.

If the flammable liquid is water soluble, such as alcohols, then addition of sufficient water will dilute the liquid to the point where it is no longer flammable. However, if there is a deep pool of alcohol rather than a shallow spill, the time required to obtain sufficient dilution might be so great that an aqueous foam would be a better extinguishing agent. If the nature of a liquid is unknown, an aqueous foam might be chosen instead of direct application of water.

Another important application of aqueous foam agents is on liquids or solids that are burning in spaces difficult to access, such as a room in a basement or the hold of a ship. The foam is used to flood the compartment completely.

Fire-fighting foam is a mass of bubbles formed by various methods from aqueous solutions of specially formulated foaming agents. Because foam is much lighter than any flammable liquid, it floats on the liquid, producing an air-excluding, cooling, continuous layer of vapor-sealing, water-bearing material that can halt or prevent combustion.

Fire-fighting foams are formulated in several ways for fire extinguishing action. Some foams are thick and viscous, forming tough heat-resistant blankets over burning liquid surfaces and vertical areas. Other foams are thinner and spread more rapidly. Some are capable of producing a vapor-sealing film of surface-active water solution on a liquid surface, and some are meant to be used as large volumes of wet gas cells for inundating surfaces and filling cavities. Various methods for applying foams are used.

The use of foam for fire protection requires attention to its general characteristics. Foam breaks down and vaporizes its water content under attack by heat and flame. Therefore, it must be applied to a burning surface in sufficient volume and rate to compensate for this loss and to provide an additional amount to guarantee a residual foam layer over the extinguished portion of the burning liquid. Before starting to apply foam to a large fire, the first step is to accumulate a sufficient number of barrels of foam concentrate to do the job. Nothing will be accomplished by putting out only part of a fire and then running out of foam, because the fire will build back to its original intensity.

Foam is an unstable air–water emulsion and can be broken down easily by physical or mechanical forces. Certain chemical vapors or fluids can destroy foam quickly. When certain other extinguishing agents are used in conjunction with foam, severe breakdown of the foam can occur. Turbulent air or violently uprising combustion gases can divert light foam from the burning area.

Protein Foaming Agents

Protein-type air foams are generated from aqueous liquid concentrates that are mixed proportionally with water. Although proportions will vary, in a typical case either 3 percent or 6 percent of concentrate will be mixed with either freshwater or seawater, with sufficient air added to cause an 8 to 1 expansion.

The concentrate contains high-molecular-weight natural protein polymers derived from a chemical digestion and hydrolysis of natural protein solids. The polymers impart elasticity, mechanical strength, and water retention capability to the foams. The concentrates also contain dissolved polyvalent metallic salts that aid in bubble strength when the foam is exposed to heat and flame. Organic solvents are added to the concentrate to improve foaming and to control viscosity at lower temperatures. These protein-type concentrates produce dense, viscous foams that have high stability and high heat resistance. In addition, they are biodegradable.

One method of evaluating these and other types of foams is to fill a graduated cylinder with the foam and then measure the time required for a certain fraction of the water to drain to the bottom. The more stable the foam, the slower it will drain.

Fluoroprotein Foaming Agents

Fluoroprotein foam concentrates contain, in addition to natural protein polymers, fluorinated surface-active agents. Fluoroprotein foams are more effective than some other foam agents for deep-liquid hydrocarbon fires because they confer a "fuel-shedding" property to the foam generated. This property is beneficial under fire-fighting conditions where the foam becomes coated with fuel, such as during subsurface injection of foam for tank fire fighting, or in certain nozzle-foam or monitor-foam applications. In addition, these foams demonstrate better compatibility with dry chemical agents than do the protein-type foams. Either 3 percent or 6 percent of the concentrate by volume can be mixed with either freshwater or seawater.

Aqueous Film-Forming Foaming Agents

These foaming agents consist entirely of synthetic materials. They form foams with water and air that are similar to the protein foams; however, in addition, they form aqueous films on the surface of flammable liquids. *Aqueous film-forming foam (AFFF)* is used at either 3 percent or 6 percent concentration by volume, mixed with either freshwater or seawater.

The air foams generated from AFFF solutions possess low viscosity, have fast spreading and leveling characteristics, and act as surface barriers to exclude air and halt fuel vaporization just as other foams do. These foams also develop a continuous layer of solution under the foam, which helps to seal and cool the flammable-liquid surface. The film, which can spread over flammable liquid surfaces not fully covered with foam, is self-healing following mechanical disruption, as long as a reservoir of foam remains nearby.

The film will spread over the surface only when the surface tension of the combustible liquid is greater than the surface tension of the AFFF solution by an amount in excess of the interfacial tension between the two liquids. Typically, surface tensions of hydrocarbon liquids and AFFF solutions are 24 and 16 dyn/cm, respectively, while the interfacial tension between the two liquids is between 1 and 6 dyn/cm, so the criterion is met. (Ordinary protein-type foam solutions have surface tensions of about 45 dyn/cm, so film-spreading of protein foam solutions over a hydrocarbon surface cannot occur; however, the foam itself can spread.)

The surface-active agent in AFFF foams is a molecule that can be thought of as having a "head" and a "tail." The head end is highly soluble in water; the tail consists of a fluorocarbon chain, such as [$CF_3CF_2CF_2CF_2CF_2CF_2CF_2$—], which is less soluble in water and more soluble in the hydrocarbon. These molecules then concentrate themselves in the interface between the two liquids, as well as the interface between the AFFF solution and air, and have a large effect on interfacial tension values.

AFFF can be used to protect flammable liquids that have not become ignited. Because of the extremely low surface tension of the solution draining from AFFF, these foams can be used in situations where deep penetration is needed, in addition to the surface-spreading action of foam itself.

Foam-generating devices yielding homogeneous foams are not necessarily needed when AFFF is used. Unlike most other foaming agents, AFFF requires less sophisticated foaming devices, such as water-spray nozzles and sprinklers, because of the inherent rapid and easy foaming capability of AFFF solutions. However, such foams drain comparatively rapidly. AFFF is compatible with dry chemical agents. Although AFFF concentrates should not be used with other types of foam concentrates, AFFF foams do not break down other foams in fire-fighting operations.

Alcohol-Type Foaming Agents

Air foams generated from ordinary agents are subject to rapid breakdown and loss of effectiveness when used on burning combustible liquids that are water-soluble or polar in nature, such as *alcohols*. Special foaming agents have been developed that produce foams suitable for application to spill or deep-liquid fires of either hydrocarbon or water-soluble (polar) flammable liquids. The most common such agents are proprietary compositions usually described as polymeric alcohol-resistant AFFF concentrates. *(See Figure 14.4.)* They exhibit AFFF characteristics on hydrocarbons and produce a floating gellike mass for foam buildup on water-soluble flammable liquids, such as alcohols.

FIGURE 14.4 Fire in water-soluble liquids being extinguished by dilution with water or by application of alcohol-type foams, which resist the breakdown that occurs with regular air foams.

Burning combustible liquids that require alcohol-type foaming agents for extinguishment include alcohols, glycols, acetone, methyl ethyl ketone, isopropyl ether, acrylonitrile, ethyl acetate, the amines, enamel thinners, and lacquer thinners. Even small amounts of these substances mixed with hydrocarbons (e.g., gasohol) cause a rapid breakdown of ordinary fire-fighting foams.

Medium- and High-Expansion Foaming Agents

These foams are suited particularly well for flooding of confined spaces. *(See Figure 14.5.)* Foams of expansions from 20 to 1 up to 1000 to 1 can be generated. They are not well suited for outdoor use because wind may blow them away from burning liquids.

Liquid concentrates for producing these foams consist of synthetic surfactants, generally 2 percent in water. Medium-expansion foam also can be generated from 3 percent or 6 percent solutions of fluoroprotein, protein, or AFFF concentrates. The effectiveness of this type of foam includes the following factors:

1. It prevents air from reaching the fire.
2. It generates steam, which dilutes the air as well as absorb heat.
3. It penetrates crevices because of low surface tension.
4. It provides protection of exposed materials that are not yet burning.

A 500 to 1 foam can be used to control fires and reduce vaporization from *liquefied natural gas* (LNG) spills.

FIGURE 14.5 High-expansion foam being applied in an aircraft hangar. Courtesy of Factory Mutual Research Corporation.

When air from inside a burning building is used to generate high-expansion foam, the presence of combustion and pyrolysis products or elevated temperature has an adverse effect on the volume and stability of the foam produced. However, this effect can be compensated for by using higher rates of foam generation.

Anyone who enters a foam-filled space should use a self-contained breathing apparatus (SCBA). Also, vision and hearing are drastically reduced in a foam-filled space, and use of a lifeline is advisable.

Techniques for Application of Foam

A large variety of both portable and in-place devices for generating and delivering foam are available commercially. These devices are described by Scheffey. [4] Figure 14.6 shows the elements of a foam extinguishing system. Another device worthy of mention is called a monitor, which is a foam cannon capable of projecting foam as far as 100 m (300 ft), in the absence of wind. However, high-expansion foam cannot be projected this way, but must approach the fire by flowing over a surface. (*See Figure 14.5.*)

Inert Gases

Water's action to extinguish fires is primarily by cooling, although the formation of steam helps to dilute the concentration of oxygen. On the other hand, inert gases act to extinguish a fire primarily by dilution. Carbon dioxide is the

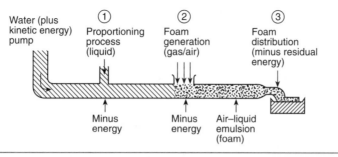

FIGURE 14.6 The steps in air-foam generation. From Scheffey. [4]

most commonly used inert gas, although nitrogen or steam could be used. Theoretically, helium, neon, or argon could be used, but they are expensive and there is no reason to use these gases except in certain special cases, such as magnesium fires.

Table 14.2 presents the minimum proportions of carbon dioxide or nitrogen gas which, if added to air, will form an atmosphere in which various vapors will not burn. On a volume basis, carbon dioxide is substantially more effective than nitrogen. However, note that a given volume of carbon dioxide is 1.57 times as heavy as nitrogen (44/28 molecular weight ratio), so the two gases have more nearly equal effectiveness on a weight basis. Either gas, in sufficient quantity, will prevent the combustion of anything except certain metals and certain unstable chemicals, such as pyrotechnics, solid rocket propellants, or hydrazine. (Unstable chemicals are discussed in Chapter 15.)

TABLE 14.2 Minimum required volume ratios of carbon dioxide or nitrogen to air that will prevent burning of various vapors at 25°C

Vapor	Carbon dioxide		Nitrogen	
	CO_2/air	% O_2	Extra N_2/air	% O_2
Carbon disulfide	1.59	8.1	3.0	5.2
Hydrogen	1.54	8.2	3.1	5.1
Ethylene	0.68	12.5	1.00	10.5
Ethyl ether	0.51	13.9	0.97	10.6
Ethanol	0.48	14.2	0.86	11.3
Propane	0.41	14.9	0.78	11.8
Acetone	0.41	14.9	0.75	12.0
n-Hexane	0.40	15.0	0.72	12.2
Benzene	0.40	15.0	0.82	11.5
Methane	0.33	15.7	0.63	12.9

Calculated from tabulations by Kuchta, p. 31. [5] Additional data are available in Zabetakis. [6]

Steam, if available, also can be used as an inert extinguishing agent. The percentage by volume required is intermediate between that required for carbon dioxide and for nitrogen.

Table 14.2 shows that the required addition of either carbon dioxide or nitrogen reduces the oxygen level to a point where exposed humans will suffer undesirable effects. (See Chapter 11, "Combustion Products.") In the case of carbon dioxide, an additional serious physiological effect, an increase in the rate and depth of breathing, will occur at the concentrations required to extinguish a fire. (See Chapter 11, "Combustion Products.")

Table 14.2 refers only to vapors, but the data are relevant to liquids or solids because they burn only by vaporizing or pyrolyzing. Accordingly, application of an inert gas can extinguish the flame over a liquid or solid. However, if the inert gas dissipates after several minutes, because, for example, the enclosure is not air-tight, it is possible that the fire could reignite from a glowing ember or hot metal. Reignition is common for a deep-seated fire, such as in upholstered furniture or in a carton of documents.

Some explanation of the physical forms of carbon dioxide is appropriate. Carbon dioxide is unusual in that it can exist only as a gas or solid at normal atmospheric pressure, but not as a liquid. Figure 14.7 shows the phase diagram of carbon dioxide.

The solid form of carbon dioxide, commonly known as *dry ice*, at atmospheric pressure, exists only below –79°C (–110°F), at which temperature it undergoes sublimation directly to the vapor, without melting. However, carbon dioxide liquid can exist at elevated pressures, as long as the temperature is above –57°C (–70°F) and the pressure is above 5.2 atm. This temperature and pressure condition is known as the *triple point* of carbon dioxide because it is the only condition at which solid, liquid, and vapor can co-exist.

Liquid carbon dioxide can be kept in a pressure vessel at any temperature between –57°C and +31°C (the *critical temperature*). Above the critical temperature, there will no longer be a liquid-gas interface in the pressure vessel, so the fluid in the vessel will be a gas. A pressure vessel at 21°C (70°F) containing liquid carbon dioxide would be at a pressure of 58 atm, which is the vapor pressure of carbon dioxide at that temperature. This pressure is used to expel liquid carbon dioxide from a cylinder in fire fighting. The cylinder normally will contain an internal dip tube reaching to the bottom so that liquid rather than vapor will be discharged. As the liquid droplets emerge from a nozzle into the lower-pressure environment, instantaneous evaporation occurs, with evaporative cooling of the residual liquid in each drop. This process causes solidification of the residual portion into dry ice particles at –79°C. If the liquid was originally at 21°C, about 75 percent of the discharged liquid would have evaporated and about 25 percent would have been converted to dry ice particles.

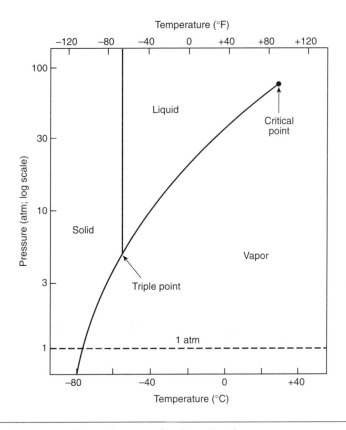

FIGURE 14.7 Phase diagram of carbon dioxide.

Some of the dry ice particles might impinge on a combustible surface and have a cooling effect; however, because the heat of sublimation of carbon dioxide is only about $1/4$ the heat of vaporization of water and because only about $1/4$ of the carbon dioxide discharged is converted to dry ice, the cooling effect on a hot surface is only about $1/16$ that produced by water discharged at an equal rate (on a mass basis).

The advantage of using carbon dioxide or nitrogen instead of water is that water damage is avoided, and there is no need to remove the agent after the fire. However, there are disadvantages to using carbon dioxide or nitrogen:

1. The toxicity problem requires that all personnel not equipped with self-contained breathing apparatus are evacuated before flooding a compartment with inert gas.

2. The fire could reignite if the inert gas dissipates before all embers or metals have cooled.

3. It is not practical to flood a really large compartment, such as a warehouse or aircraft hangar; and, obviously, flooding is not applicable to a fire in the open.

4. Inert gas available from a cylinder is limited in quantity; the quantity of water taken from a public water supply is virtually unlimited. A supply of inert gas could become exhausted, particularly in the case of multiple reignitions.

5. It is not possible to project a gas very far from a nozzle, as it is with water.

Carbon dioxide, of course, can be used conveniently as the agent in a hand-held extinguisher to combat a very small fire that can be approached closely. This provides a clean way of extinguishing a small fire that is not deep-seated. Fires in high-voltage electrical equipment also can be fought safely in this way. In addition, an inert gas can be used to flood a small electronics-containing cabinet in a large room or a restaurant kitchen range. *(See Figure 14.8.)*

FIGURE 14.8 A carbon dioxide extinguishing system for a restaurant kitchen range and exhaust system. Outlets are arranged so that the carbon dioxide under the hood floods the surfaces of the cooking equipment and enters the exhaust system to extinguish grease fires in the duct system. From the *Fire Protection Handbook*. [7]

To compare carbon dioxide and nitrogen: Carbon dioxide has the advantage that it can be stored in a cylinder as a liquid at a relatively moderate pressure of 58 atm (21°C), while nitrogen at the same temperature must be stored as a gas, usually at about 140 atm. A given size of cylinder at 21°C at these pressures could hold about 3 times as large a volume of carbon dioxide as nitrogen (measured after expansion, at atmospheric conditions). Nitrogen also can be stored for short periods more compactly as a cryogenic liquid at –196°C and 1 atm; however, long-term storage would result in continuous loss of nitrogen. As a result of these factors, carbon dioxide is used more commonly than nitrogen as an inerting gas.

Semipermeable membranes that can inexpensively separate the oxygen and nitrogen of the air are being developed. When and if such systems become cost effective, it might be practical to provide permanent nitrogen-inerting of hazardous spaces that do not require human presence. A reduction of the oxygen percentage in the air from 21 percent to 10 percent by volume would make fires and explosions impossible, except for a few special gases, such as hydrogen, acetylene, or carbon disulfide, which would require greater dilution.

Halogenated Agents

Halogenated extinguishing agents are chemical derivatives of simple hydrocarbons, such as methane (CH_4) or ethane (CH_3——CH_3), in which some or all of the hydrogen atoms have been replaced with fluorine, chlorine, or bromine atoms, or by some combination of these halogen elements. These agents are liquids when stored in pressurized tanks at normal temperatures, but most of them are gases at atmospheric pressure and normal temperatures. The most effective agents contain bromine.

The halogenated agents can be used for fire applications such as those discussed previously for carbon dioxide (e.g., electrical fires, cases where water or dry chemicals would cause damage, or inert-gas flooding of compartments). Halogenated agents have three principal advantages over carbon dioxide:

1. Certain halogenated agents are significantly more effective than carbon dioxide, either on a weight or a volume basis. (However, they are more costly.)

2. Certain halogenated agents are effective in such low volumetric concentrations that sufficient oxygen remains in the air after compartment flooding for comfortable breathing.

3. For several halogenated agents, only partial vaporization occurs initially during projection from a nozzle, and the liquid can be projected farther than carbon dioxide.

The drawbacks of using halogenated agents are discussed later in this subsection; they involve the following:

1. Toxicity and corrosivity of their decomposition products
2. Detrimental effect of halogenated compounds on earth's ozone layer

Historically, carbon tetrachloride (CCl_4), methyl bromide (CH_3Br), and chlorobromomethane (CH_2ClBr) were used as extinguishing agents, but they were banned from use in the 1950s because of the toxicity not only of the decomposition products but primarily of the compounds themselves. They were replaced by more recently developed agents that incorporated fluorine, bromine, and in one case chlorine. These new agents are substantially less toxic than the formerly used agents, and they also decompose less readily at high temperatures. Furthermore, they are effective as inerting agents in much lower concentrations.

A halon system of nomenclature was introduced for these compounds. A four-digit number is used for each compound, each digit having significance as follows:

First digit: number of carbon atoms

Second digit: number of fluorine atoms

Third digit: number of chlorine atoms

Fourth digit: number of bromine atoms

If the fourth digit is zero, it is omitted, and only three digits are used.

For example, Halon 1301 has 1 carbon atom, 3 fluorine atoms, 0 chlorine atoms, and 1 bromine atom. It is called bromotrifluoromethane. Halon 104 (carbon tetrachloride) has 1 carbon atom, 0 fluorine atoms, and 4 chlorine atoms per molecule. Halon 2402 (dibromotetrafluoroethane) has 2 carbon atoms, 4 fluorine atoms, 0 chlorine atoms, and 2 bromine atoms.

Of the various halons, Halon 1301 has the lowest toxicity, as well as the highest effectiveness on a weight basis. Among the highly effective halons, it has the highest volatility, which is desirable for flooding applications. However, if a halon liquid is needed for direct application to a burning surface, to accomplish cooling as well as inerting of the nearby region, then a less volatile halon, such as Halon 1211 (bromochlorodifluoromethane) or Halon 2402 would be preferred.

Table 14.3 gives the physical properties of Halons 1301, 1211, and 2402. They are all liquids at normal temperatures when stored in pressurized tanks. They can be stored under high-pressure nitrogen if the liquid must be expelled from the tank more rapidly than under the vapor pressure of the halon alone. The use of nitrogen for pressurization is especially important for outdoor storage in the winter.

TABLE 14.3 Physical properties and chemical formulas of three halon extinguishing agents

Extinguishing agent	Halon 1301 CF_3Br	Halon 1211 CF_2ClBr	Halon 2402 $C_2F_4Br_2$
Boiling point (°C)	−58	−4	+47
Liquid density at 20°C (g/cc)	1.57	1.83	2.17
Latent heat of vaporization (J/g)	117	134	105
Vapor pressure at 20°C (atm)	14.5	2.5	0.46

The inerting capabilities of Halon 1301 and Halon 1211 are shown in Table 14.4. If methane in any proportion is combined with a mixture containing 5.4 volumes of Halon 1301 and 100 volumes of air, at 25°C, then no combustion can result. By contrast, a mixture containing 33 volumes of carbon dioxide and 100 volumes of air would have been required to obtain the same result. This suggests that a molecule of Halon 1301 is 33/5.4 = 6.1 times as effective as a molecule of carbon dioxide. Note, however, that the molecular weight of Halon 1301 is 149, while that of carbon dioxide is 44, so the ratio of molecular weights is 149/44 = 3.39. Accordingly, on a weight basis, Halon 1301 is only 6.1/3.39 = 1.8 times as effective as carbon dioxide for methane fires.

Table 14.4 shows that the inerting proportion of halon needed varies somewhat depending on the nature of the combustible, and substantially more halon is needed for hydrogen, carbon disulfide, or ethylene fires than for most other combustibles. Table 14.4 also shows that Halon 1301 and Halon 1211 have similar effectiveness on a volume basis for most combustibles. On the basis of molecular weights, a molecule of Halon 1211 is 165.5/149 = 1.11 times as heavy as a molecule of Halon 1301.

Table 14.4 is based on experiments in which a strong ignition source is applied to a uniform combustible-air–halon mixture and the occurrence or nonoccurrence of a propagating flame is noted. A somewhat smaller quantity of halon would be needed to cause an existing flame on a burner to become unstable and then extinguish itself; the quantity of halon needed will depend on the details of the burner. A flame burning on a solid is even more easily extinguished. Taylor [8] stated that the flames on most solids, other than with deep-seated fires, could be extinguished with 4 percent to 6 percent by volume of Halon 1301 in the surrounding atmosphere.

The subsection entitled Inert Gases addressed problems with extinguishment of deep-seated fires by carbon dioxide or nitrogen; this is equally valid when a halon agent is used. Unless a sufficient quantity of the halon in liquid form can reach the "seat of the fire" and cool all the solid sufficiently, reignition can occur after the agent has dissipated. If the halon reaches the combustible as a gas via compartment-flooding, then no such cooling can occur, and

TABLE 14.4 Minimum required volume ratios of halons to air at 25°C that will prevent burning of various vapors

Vapor	Halon 1301		Halon 1211	
	1301/air	% O_2	1211/air	% O_2
Hydrogen	0.29	16.2	0.43	14.7
Carbon disulfide	0.15	18.2	—	—
Ethylene	0.13	18.5	0.114	18.8
Propane	0.073	19.5	0.065	19.7
n-Hexane	—	—	0.064	19.7
Ethyl ether	0.070	19.6	—	—
Acetone	0.059	19.8	0.054	19.9
Methane	0.054	19.9	0.062	19.7
Benzene	0.046	20.0	0.052	19.9
Ethanol	0.045	20.0	—	—

Calculated from tabulations by Kuchta, p. 32. [5]

the effect of the halon is to extinguish the gaseous flame without affecting the pyrolysis or smoldering, with the possibility of later resumption of flaming.

Table 14.4 also shows that the addition of Halon 1301 or 1211 needed in the air will only reduce the oxygen percentage from 21 percent to about 19 percent for most combustibles, while the required amount of carbon dioxide would have reduced the oxygen level to 14 percent or 15 percent. Furthermore, the physiological effects of carbon dioxide on humans (at the concentrations needed for inerting) are greater than those of Halon 1301.

According to Taylor [8], after 30 years of experience with Halon 1301, no significant adverse health effects have been reported from the proper use of the agent as an extinguishant. Exposure of animals to concentrations of Halon 1301 in air greater than 30 percent by volume has caused convulsions, lethargy, and unconsciousness. These effects reportedly disappear after removal from the exposure. Human subjects have been exposed to concentrations in air less than 7 percent by volume with little effect. Above 7 percent, tingling of the extremities and dizziness have been reported. Above 10 percent, pronounced dizziness and reduction of dexterity have been observed. These effects disappear quickly after removal from the exposure. Because 5 percent by volume is the usual design concentration for inerting a compartment, toxicity usually is not a serious concern. NFPA 12A [9] permits Halon 1301 design concentrations up to 10 percent in normally occupied areas. [8]

When Halon 1301 is applied to a fire, a certain proportion of the agent that contacts flame or hot surfaces will decompose, producing hydrogen fluoride and hydrogen bromide, as well as minuscule traces of other halogen compounds.

On the basis of toxicity data for hydrogen fluoride and hydrogen bromide [10], and assuming an initial 5 percent concentration of Halon 1301 in a given volume, the following calculation can be made. If 1 percent of the Halon 1301 decomposes to form HF and HBr, and the gas remains undiluted by additional air, then a 15-minute exposure to this gas would be lethal. However, the following can be argued:

1. The sharp acrid odor of HF and HBr would warn any compartment occupants that the gas was present.

2. Under these conditions, other lethal toxicants from the fire would also be present.

Note that very small concentrations of HF or HBr can produce corrosion on metals. (See Chapter 11, "Combustion Products.") Further, halons should not be used on metal fires, because of exothermic reactions between halons and metals.

It has been found that halons and other chemicals work their way into earth's upper atmosphere, where they appear to act as catalysts for the conversion of ozone, O_3, to normal oxygen, O_2. [11] The ozone in the upper atmosphere plays a valuable role in filtering out the far-ultraviolet radiation of the sun, which would otherwise do damage to plant and animal life on earth's surface. Destruction of the ozone layer also might affect the world's weather. Accordingly, there has been international activity directed toward eliminating the production or release of halons into the atmosphere. In 1994 production of fire protection halons ceased in the United States and in most other countries. Current stockpiles of halon are still being used, but a shift is underway to alternatives wherever possible.

It would, of course, be highly desirable if chemists discovered a substitute for Halon 1301 or Halon 1211 that provided both inerting and breathability qualities and did not attack the ozone layer. In considering this possibility, a review follows of what is known about why the CF_3Br and CF_2ClBr molecules are so much more effective than the carbon dioxide molecule.

Figure 14.9 shows a methane–air flammability limit diagram as modified by various volumetric proportions of Halons 1301 and 1211 or carbon dioxide. The enormous difference is obvious.

It has been established that carbon dioxide acts by absorbing heat and reducing the flame temperature from about 2200 K for a stoichiometric fuel–air mixture to about 1500 K to 1600 K; below these temperatures, most flames can no longer burn. (Hydrogen, acetylene, and carbon disulfide flames, however, are exceptions; they continue to burn at even lower temperatures.)

If nitrogen were added to a stoichiometric fuel–air mixture instead of carbon dioxide, a somewhat larger volume of inert agent would be needed, because the heat capacity of the nitrogen molecule is less than that of the car-

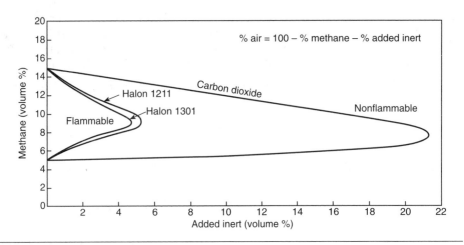

FIGURE 14.9 Flammability limits for methane–air mixtures with added inerting agents. Source: Kuchta, p. 30. [5]

bon dioxide molecule. Similarly, if argon, which has an even lower heat capacity per molecule, were added, an even larger volume of argon would be needed for inerting. In each case, the flame would go out when the temperature dropped below 1500 K to 1600 K.

However, if a small proportion of bromotrifluoromethane were added to a flame, so that the temperature dropped to only about 1800 K, the flame would be extinguished. Clearly, the mechanism is different.

The most important chemical reactions in flames involve the free atoms H and O and the free radical OH, which undergo chain reactions with the fuel and oxygen. Particularly, the branching chain reaction $H + O_2 \rightarrow OH + O$ is very important. It is believed that the CF_3Br molecule decomposes in the flame to form HBr, and the HBr then acts to remove H atoms and OH radicals, by the following two combustion reactions:

$$HBr + H \rightarrow H_2 + Br$$

and

$$HBr + OH \rightarrow H_2O + Br$$

It happens that the molecules HF and HCl cannot react as rapidly with H or OH as can HBr, so bromine appears to be essential to the inerting molecule. It has been found that hydrogen iodide, HI, is about as effective as HBr, but iodine is more expensive and heavier than bromine, as well as quite toxic, so iodine appears to offer no advantage over bromine for flame extinguishment.

In addition to the destruction of chain carriers, it has been speculated that a secondary contribution of halons to flame extinguishment comes from the extreme sootiness of halogen-containing flames. The more sooty (luminous) the flame, the greater the radiative heat loss and the lower the temperature.

The role of fluorine in halogenated agents is twofold:

1. The fluorine atoms replace hydrogen atoms in the methane or ethane, thereby reducing the flammability of the inerting agent itself.
2. The toxicity of the agent is reduced.

For example, CH_3Br is much more toxic than CF_3Br, and, again, CH_2ClBr is much more toxic than CF_2ClBr.

This current degree of understanding about why CF_3Br is such a good inerting agent for flames does not provide clear guidance about how other equally effective gaseous agents not containing bromine could be found. It appears that any volatile compound containing bromine, upon reaching the upper atmosphere, would destroy ozone.

The vast majority of known molecules are liquids or solids, not gases, at room temperature and 1 atm. All known gaseous molecules have been considered for inerting effectiveness, but no practical substitute as effective as CF_3Br has emerged.

In 1973 Huggett [12] indicated that completely fluorinated compounds of carbon, such as C_2F_6 or C_3F_8, should be considered as inerting agents. His measurements showed that kerosene could not be ignited in an atmosphere consisting of 7 percent C_3F_8 in air, or 8 percent C_2F_6 in air. He reported data suggesting that the toxicity of these perfluorocarbons is minimal. Currently, perfluorobutane, C_4F_{10}; heptafluoropropane, C_3HF_7; trifluoromethane, CHF_3; and pentafluoroethane, C_2HF_5 are being studied as replacement agents for Halon 1301. Tests show that all are effective in extinguishing flames, but they are substantially less effective than Halon 1301 on either a mass or a volume basis.

Another class of possible replacement agents is the hydrochlorofluorocarbons. An example is chlorotetrafluoroethane, C_2HF_4Cl. This also is effective, but much less so than Halon 1301. (See DiNenno [13] for more information on halon replacements.)

At present it appears that the choices for halon replacement are among one of the types of compounds mentioned above, which compromise the halon values of weight, volume, and breathable oxygen percentage in the inerted air. However, another type of extinguishing agent, possibly a water mist (which is not yet fully proved), might be considered.

Dry Chemical Agents

Dry chemical powders provide an alternative to carbon dioxide or the halons for extinguishing a fire without the use of water. These powders, which are 10 to

75 μm in size, are projected by an inert gas. Of the seven types of dry chemicals in use, only one of these, monoammonium phosphate, is effective against deep-seated fires because of a glassy phosphoric acid coating that forms over the combustible surface. All forms of dry chemical act to suppress the flame of a fire.

One reason that dry chemical agents other than monoammonium phosphate are popular has to do with corrosion. Any chemical powder can produce some degree of corrosion or other damage, but monoammonium phosphate is acidic and corrodes more readily than other dry chemicals, which are neutral or mildly alkaline. Furthermore, corrosion by the other dry chemicals is stopped by a moderately dry atmosphere, while phosphoric acid has such a strong affinity for water that an exceedingly dry atmosphere would be needed to stop corrosion.

Application of any dry chemical agent on electrical fires is safe (from the viewpoint of electric shock) for fire fighters. However, these agents, especially monoammonium phosphate, can damage delicate electric equipment.

The most common application of dry chemical agents is for relatively small flammable liquid fires. For the special case of kitchen fires involving hot fat, monoammonium phosphate is not recommended because of its acidic nature; an alkaline dry chemical, such as potassium bicarbonate, is preferred.

Table 14.5 lists the chemical names, formulas, and popular or commercial names of the various dry chemical agents. In each case, the particles of powder are coated with an agent, such as zinc stearate or a silicone, to prevent caking and promote flowing. The effectiveness of any of these agents depends on the particle size. The smaller the particles, the less agent is needed as long as particles are larger than a critical size. [14] The reason for this fact is believed to be that the agent must vaporize rapidly in the flame to be effective. [15] However, if extremely fine agent were used, it would be difficult to disperse and apply to the fire.

Precise comparisons of effectiveness of one dry chemical with another are difficult to make because a comparison to reveal chemical differences would require each agent to have identical particle size, which is difficult to achieve. Furthermore, gaseous agents can be compared by studying flammability limits of uniform mixtures at rest; however, if particles were present, they would settle out unless the mixture was agitated, thus modifying the combustion behavior. Nevertheless some comparisons of various powders have been made. The results generally show the following:

1. Sodium bicarbonate and sodium chloride have comparable effectiveness and are several times as effective (on a weight basis) as powders such as limestone or talc, which are supposedly chemically inert in a flame.

2. Potassium bicarbonate or potassium chloride is up to twice as effective (on a weight basis) as the corresponding sodium compounds.

TABLE 14.5 Dry chemical agents

Chemical name	Formula	Popular name(s)
Sodium bicarbonate	$NaHCO_3$	Baking soda
Sodium chloride	$NaCl$	Common salt
Potassium bicarbonate	$KHCO_3$	"Purple K"
Potassium chloride	KCl	"Super K"
Potassium sulfate	K_2SO_4	"Karate Massiv"
Monoammonium phosphate	$(NH_4)H_2PO_4$	"ABC" or multipurpose
Urea + potassium bicarbonate	$NH_2CONH_2 + KHCO_3$	"Monnex"

3. Under some conditions, monoammonium phosphate is more effective than potassium bicarbonate [16], while under other conditions it is claimed to be less effective.

4. Monnex has been claimed to be twice as effective as potassium bicarbonate because of the rapid thermal decomposition of the complex formed between urea and potassium bicarbonate, causing a breakup of the particles in the flame to form very fine fragments, which then gasify rapidly.

It seems clear that the effective powders act on a flame by some chemical mechanism, presumably forming volatile species that react with hydrogen atoms or hydroxyl radicals. However, the precise reactions have not been established firmly. Although the primary action is probably removal of active species, the powders also discourage combustion by absorbing heat, by blocking radiative energy transfer, and, in the case of monoammonium phosphate, by forming a surface coating.

According to Hague [17], the ingredients used in dry chemical agents are nontoxic but can cause temporary breathing difficulty and can interfere with visibility.

FIRE CLASSES

For convenience in selecting agents for fire fighting, fires have been divided into four classes, A, B, C, and D, with the idea that particular agents can be recommended for particular classes of fires:

Class A: Fires in "ordinary combustible materials," such as wood, paper, cloth, and certain plastics

Class B: Fires in flammable liquids, gases, and greases

Class C: Fires involving energized electrical equipment, where a shock hazard is present

Class D: Fires in combustible metals

It must be noted that a number of the most common plastics, such as polyethylene, polypropylene, and polystyrene, form a low-viscosity melt when burning and have some of the characteristics of a flammable liquid rather than a solid. This fact limits the usefulness of the ABCD system of categorizing fires.

The British system of classifying fires differs from the U.S. system. In their system, Class A fires involve carbon compounds that form glowing embers; Class B fires involve liquids or liquefiable solids, of which B1 are miscible with water and B2 are immiscible with water; Class C fires involve flammable gases or vapors; and Class D fires involve metals. Electrical fires in energized equipment constitute a fifth category, Class E.

In contrast to the general-purpose agents discussed so far, a number of agents are available for use in special situations. For example, rock dust is commonly spread on the floor of a coal-mine gallery to suppress potential fires and explosions. For liquefied natural gas (LNG) spills, a special agent has been developed from granules of cellular glass that float on top of the spill, reducing the burning rate. Metal fires, discussed at the end of this chapter, can be fought with dry sand, powdered graphite, carbon microspheres, boron trifluoride, helium, argon, dry sodium chloride, dry sodium carbonate, or any of several proprietary mixtures. The choice of agent would depend on the type of metal, the size of the fire, and the circumstances.

EXTINGUISHMENT OF GASEOUS FLAMES

Extinguishment of a fire involving a continuously flowing combustible gas often is very difficult. The best tactic is to shut off the flow of gas. If extinguishment is accomplished while the gas is still flowing inside a building, then the danger of filling the building with an explosive gas mixture is introduced. In some cases, it might be preferable to let the flame continue to burn if the flow of combustible gas cannot be stopped.

If it is not possible to shut off the gas supply, then fire fighting by any of several techniques is possible. One approach is to attack the base of the flame with a dry chemical nozzle, or carbon dioxide, or steam, or a halon. Whichever agent is used, it should be projected in the same general direction as the burning jet or plume.

When this tactic is used, it is advisable to cool any hot metal in the vicinity and to remove or deenergize any other ignition sources before attacking the fire itself. Otherwise, reignition is likely to occur after temporary extinguishment; the supply of agent could be depleted by that time.

Another approach is to interrupt the fresh air supply to the fire. If the compartment can be almost entirely closed, leaving an exhaust vent to prevent pressure buildup, then the combustion products will accumulate and dilute the air to the point where extinguishment occurs. Alternatively, if the entire compartment can be flooded with an inert agent, extinguishment will result. Again, removal of reignition sources is recommended. Table 14.6 shows the parts by volume of added suppressant gas that must be mixed with 100 parts of air by volume to suppress a methane (or natural gas), a propane, or a hydrogen fire. Kuchta [5] and Zabetakis [6] provide data for extinguishing other gas fires.

Recall that, to be effective, any of these additives (other than the halons) will reduce the oxygen in the air to dangerously low levels for exposed humans. Consequently, a false alarm or a very small fire triggering an automatic inerting system could be hazardous to building occupants.

In many cases, a somewhat lower concentration of inert gas than is necessary to ensure suppression can be adequate to cause the flame to detach itself from the exit orifice and "lift off" or "blow off." Alternatively, blow-off often can be induced by a pressure wave, as from an explosive charge, in appropriate circumstances (e.g., outdoors).

In the case of a flame propagating through a duct, the propagation can be prevented from passing a given point if a *flame arrestor* is present. One type of flame arrestor, used in gas distribution pipes for welding, involves bubbling the mixture through water in discrete bubbles. Another common type causes the mixture to flow through narrow passageways, such as a wire screen, a metal honeycomb, or a porous metal plate. The metal extracts heat from the flame and quenches it, thus preventing it from passing through.

However, if the combustion has built up substantial pressure on the flame side of the arrestor, hot gases can be projected through the openings so rap-

TABLE 14.6 Amounts of suppressant gases, added to air, that will ensure noncombustibility for three combustible gases

Suppressant gas	Combustible gas (Parts by volume of suppressant gas needed per 100 parts of air by volume)		
	Methane	Propane	Hydrogen
Halon 1301 (CF$_3$Br)	3.1[a]	4.3[a]	46[b]
Halon 1211 (CF$_2$ClBr)	3.2[a]	4.8[a]	60[b]
Carbon dioxide	32	41	156
Water vapor	41	—	181[c]
Nitrogen	66	76	313
Helium	102	—	—

Data from Zabetakis [6] unless otherwise noted.
[a]Taylor. [8]
[b]Kumar et al. [18]
[c]Bajpai and Wagner. [19]

idly that they are not cooled sufficiently, and can reignite the flame on the other side. Another failure mode involves the flame continuing to burn on one side of the arrestor and gradually heating it to a temperature high enough so that it is no longer effective.

EXTINGUISHMENT OF BURNING LIQUIDS

The most common fires involving liquids are as follows:

1. Spill fires, generally characterized by shallow depth of liquid
2. Fires in open tanks or vessels, where the depth may be substantial

The flammable liquid itself has three important properties relevant to extinguishment:

3. Its fire point relative to the boiling point of water
4. Its solubility in water
5. Its density relative to water

These properties determine whether and how water can be used in fire fighting.

If a foam or a dry chemical is to be used to fight a flammable liquid fire, the compatibility of the liquid with the agent must be known. If a fire is to be attacked with a combination of several agents (e.g., foam and dry chemical), the compatibility of the agents with one another also must be considered.

If a spill fire imposes a heat load on sealed tanks or drums containing flammable liquid, it is vital that such vessels are cooled to prevent rupture and major escalation (BLEVE) of the fire. (See Chapter 9, "Fire Characteristics: Liquid Combustibles," and Figure 9.2.)

If a fire is large and must be fought from a distance, then the range of the equipment used to project the agent becomes critical. In addition, if a large fire is to be fought with an agent in limited supply, an adequate quantity of the agent must be accumulated before application is begun.

In view of these multiple constraints, the judgment of experienced personnel is valuable in choosing the appropriate agent and tactics for any particular case.

Shallow Spill Fire

A shallow spill fire will burn at very high intensity for a short time, perhaps less than a minute, by which time the liquid will be consumed and the fire will die down, unless other combustibles have become ignited. Obviously, an extinguishing agent must be applied very rapidly in order to put out the fire. An automatic system for dispersing the agent, which might be a foam, a dry chemical, or a water spray, would be most effective for this type of fire.

An automatic sprinkler system producing water spray is an inexpensive and widely available protection, but it will extinguish flammable liquid fires in only three special cases:

1. The fire point of the liquid is above the boiling point of water, so application of a gentle water spray to the shallow spill can rapidly cool the liquid below its fire point.

2. The liquid is water soluble (e.g., methanol, ethanol, propanol, acetone, or ethylene glycol), so water application can rapidly dilute a shallow spill, rendering it nonflammable.

3. The liquid is denser than water (e.g., carbon disulfide, phenol, or chlorine-containing liquids), so that the water spray can form a layer over the surface.

However, even if none of these conditions is met, a sprinkler spray, while unable to extinguish the fire, might be effective in limiting the fire damage by cooling and wetting down nearby solid combustibles, as well as the ceiling. However, unsuccessful application of sprinkler water to a flammable liquid spill will probably cause the burning flammable liquid to spread, unless a dike is present. Another problem is encountered if the liquid has a high fire point and is less dense than water. In this case, water droplets, even if applied gently, will sink below the surface and turn into steam, causing eruption of flammable liquid into the flames and increasing the burning rate. If this situation can be tolerated for a short time, the shallow spill will become cooled and the fire will go out.

Deep Tank Fire

A fire in a deep open tank differs from a shallow spill fire in that it could burn for hours or days if not extinguished. It also differs in that the depth of the liquid makes it difficult to cool or dilute.

It would be highly desirable if dry chemical or carbon dioxide could be applied in such a way that the entire flame would be extinguished in one application. Extinguishing 90 percent of the flame accomplishes nothing because reignition will occur unless the fire is completely extinguished. The various foams are ideal agents for extinguishing such fires. (See subsection entitled Aqueous Foams earlier in this chapter.) Rules exist for the required quantities and rates of application of foam needed, depending on the exposed surface area of the liquid. [20] Nearby hot metal must be cooled to prevent reignition. (Review the boilover phenomenon described in Chapter 9, "Fire Characteristics: Liquid Combustibles.")

It is extremely difficult to extinguish a flammable liquid that is burning while issuing from an opening or overflowing a tank and falling some distance to the floor. This type of three-dimensional fire (or a flammable liquid spray fire) usually is put out only by stopping the flow. The use of dikes is recommended strongly in areas where flammable liquid spills are possible, in order to limit the size of a potential fire.

EXTINGUISHMENT OF BURNING SOLIDS

For a solid to burn, a portion of it must be at a temperature high enough that pyrolysis occurs at a rate sufficient to maintain the flame. For most solids, this temperature is 300°C to 400°C, and the pyrolysis rate must be at least a few grams per square meter per second. If even a small amount of liquid water, with its high heat of vaporization, can reach this region, the solid can be cooled sufficiently to reduce or stop the pyrolysis, and the flame will be extinguished. Even deep-seated fires can be suppressed in this way. Accordingly, water is the obvious agent of choice for burning solids.

The two most common means of applying water are by the following:

1. A solid stream or spray from a hose
2. Spray from automatic sprinklers

The practical aspects of manual fire fighting and use of sprinklers are discussed by Thompson [21], Cote and Bugbee [22], Clark [23], and in the *Fire Protection Handbook* [24]. From a scientific viewpoint, studies reviewed by Heskestad [25] and Rasbash [26] have investigated the minimum rate of water application to a burning solid surface that will cause extinguishment.

An important paper by Magee and Reitz [27] reported that burning slabs of various plastics, horizontal and vertical, were simultaneously heated with radiant heaters and cooled with controlled water sprays. Extinguishment conditions were then determined. Figure 14.10 shows a linear relationship between the radiative heating rate and the water application rate required for extinguishment. The reciprocal of the slope of the line is found to be approximately the heat of vaporization of water, as theory would predict.

To extinguish burning Plexiglas, sufficient water must be applied to reduce the burning rate to less than about 4 g/m^2-s. Depending on the intensity of the externally imposed radiative flux (up to 18 kW/m^2), a water application rate of from 1.5 to 8 g/m^2-s was required. This is a very small application rate of water. For extinguishment with no external radiative flux, it was necessary to spray only enough water that the heat absorbed by its vaporization was 3 percent of the heat of combustion.

Somewhat similar results were obtained for three other plastics. However, application of water at very low rates caused an increase in the burning rates of polystyrene and especially polyethylene because droplets of water penetrated the molten plastic surface and then vaporized, causing combustible liq-

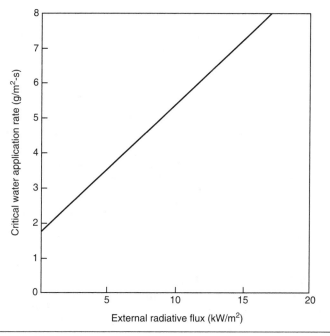

FIGURE 14.10 Water application rate needed to extinguish a fire on a vertical Plexiglas slab.

uid to be spewed into the flame. This phenomenon did not happen with the other two plastics tested — Plexiglas and Delrin — because their melts were much more viscous.

In practical fire fighting, water must be applied at 10 to 100 times the rates used in the research described above because of the difficulty of delivering the water directly to the burning surface. If a fire cannot be reached at all by water (because it is shielded in some way), then it is sometimes possible to fill the compartment with high-expansion foam and get water to the burning surface in this way.

There are three circumstances in which water might not be the agent of choice for extinguishing a burning solid combustible:

1. If there is major concern about water damage

2. If there is an electric shock hazard

3. If, for chemical reasons, water is not compatible with the burning solid

Each of these cases is discussed in greater detail, as follows.

Water Damage

To avoid water damage, the choices are to apply another extinguishing agent directly onto the fire, or to fill the compartment with an agent. In either case, the agent used is usually carbon dioxide, a halon, or a dry chemical. None of these agents, except monoammonium phosphate, are effective against deep-seated fires unless agent application is maintained long enough for cooling to occur, so that reignition is not a problem. In many cases, monoammonium phosphate can cause more damage than water.

Of course, if the agent is applied to a small fire, cooling is likely to occur sooner than for a large fire. Also, it has been found that a thin material, such as a textile, can be extinguished by potassium bicarbonate without reignition, so the use of monoammonium phosphate is not necessary in this case.

Electric Shock Hazard

When there is a potential problem of electric shock hazard, a nozzle projecting a water spray of discrete droplets is clearly less hazardous than a nozzle projecting a solid stream. Even a solid stream projecting from a 1½-in. (38-mm) nozzle orifice is not considered hazardous for voltages less than 635 V and stand-off distances more than 2.7 m (9 ft), or less than 1270 V and stand-off distances more than 4.9 m (16 ft), as long as the water is reasonably pure, that is not saltwater. [28] Still, fire fighters standing in puddles of water who touch live electric equipment are obviously at risk.

If electric power cannot be shut off, then fire fighters can avoid the risk of shock by using carbon dioxide, halon, or a dry chemical on an electrical fire.

Water Incompatibility

Water is usually the wrong agent for fires involving metals because a number of metals can react exothermically with water to form hydrogen, which, of course, burns rapidly. (See Chapter 10, "Fire Characteristics: Solid Combustibles.") Furthermore, violent steam explosions can result if water enters molten metal. As an exception, extinguishment has been accomplished when large quantities of water were applied to small quantities of burning magnesium, in the absence of pools of molten magnesium.

Table 14.7 lists extinguishing agents used for various metal fires. In general, metal fires are difficult to extinguish because of the very high temperatures involved and the correspondingly long cooling times required. Note that certain metals react exothermically with nitrogen or carbon dioxide, so the only acceptable inert gases for these metals are helium and argon. Halons should not be used on metal fires.

In addition to metals, certain inorganic chemicals are not compatible with water. For example, alkali and alkaline earth carbides, of which the best

TABLE 14.7 Extinguishing agents for metal fires

Agent	Main ingredient	Used on
Powders		
Metal Guard®	Graphite	Al, Ca, Hf, K, Li, Mg, Na, Pu, Th, Ti, U, Zr
Met-L-X®	NaCl	Al, K, Mg, Na, Ti, U, Zr
TEC® powder	KCl, NaCl, $BaCl_2$	K, Mg, Na, Pu, U
Lith-X®	Graphite	Li, Mg, Na, Zr
Na-X®	Sodium carbonate	Na
Copper powder	Cu	Al, Li, Mg
Salt	NaCl	K, Mg, Na
Soda ash	Sodium carbonate	K, Na
Gases		
Argon	Ar	Any metal
Helium	He	Any metal
Nitrogen	N_2	K, Na
Boron trifluoride	BF_3	Mg

Source: Tapscott [29].

known is calcium carbide, react with water to form acetylene, which is highly flammable. Lithium hydride, sodium hydride, or lithium aluminum hydride react with water to produce hydrogen. The peroxides of sodium, potassium, barium, and strontium react exothermically with water. Cyanide salts react with acidified water to form a highly toxic gas, hydrogen cyanide. Even if these chemicals are not combustible themselves, they could be packed in combustible cartons and thus become involved in a fire. They also could be stored in racks above combustible items.

Certain organic peroxides, used as polymerization catalysts in plastics manufacturing, are so unstable that they are stored under refrigeration to avoid exothermic heating. If water at normal room temperature were to be applied, this would provide heat to the peroxide and promote its exothermic decomposition.

A problem in applying water to fires involving toxic chemicals (e.g., pesticides) is associated with the runoff of contaminated water, which could cause groundwater pollution. In such cases, if no other agent but water is available or practical, then there might be only two alternatives:

1. To use the minimum quantity of water possible
2. To allow a building to burn, thus producing downwind air pollution instead of water pollution if the fire does not completely destroy the toxic chemicals

The variety of reactive chemicals is so great that it is not practical to present here the best techniques of fire fighting for each case. Further information can be obtained in the *Fire Protection Guide on Hazardous Materials*. [30] The Chemical Referral Center (Chemical Manufacturers Association, 2501 M St., NW, Washington D.C. 20037) will provide assistance in obtaining nonemergency safety information on chemicals. Their telephone number is 1-800-CMA-8200.

Problems

1. List four ways in which the properties of water can be modified by additives to improve fire-fighting performance.

2. Is it preferable for an automatic sprinkler to discharge small drops or large drops? Why?

3. What is the best agent for fighting a gasoline spill fire? A gasoline–alcohol mixture (gasohol) spill fire?

4. When should a high-expansion foam be used instead of a regular foam?

5. Can foam be used to fight a large fire that, because of intense thermal radiation, cannot be approached closer than 60 m?

6. By what principle are certain foams able to form aqueous films over the surface of a liquid spill?

7. A compartment containing a fire is flooded with an inert gas such as carbon dioxide, and the flame is extinguished. After 10 min, the inert atmosphere is dissipated by leaks. What could happen next?

8. Does a cylinder of carbon dioxide at room temperature contain a gas, a liquid, or a solid? When the valve is opened, what emerges?

9. What are the advantages and disadvantages of fire fighting with water instead of an inert gas?

10. Why is carbon tetrachloride no longer used as a fire-fighting agent?

11. From the viewpoint of fire fighting, where does polyethylene (the most common plastic) fit in the various classes of combustibles (A, B, C, or D)?

12. What are the relative extinguishing effectiveness qualities of an inert powder such as limestone and a chemically active powder such as potassium bicarbonate?

13. Why does fighting a liquid spill fire require knowledge of what liquid is burning?

14. What is the best way to attack a fire involving a flowing gas?

15. What is the main reason that water is so effective in extinguishing burning solids (other than metals)?

16. Under what conditions can water be applied to "live" electrical equipment without shock hazard?

17. Name some solids, in addition to metals, that are not compatible with water application in a fire situation.

References

1. Cote, A. E., ed., *Fire Protection Handbook,* 18th ed., National Fire Protection Association, Quincy, MA, 1997, Section 6, pp. 6-1 – 6-405.

2. You, H. Z., "Sprinkler Drop-Size Measurement," FMRC J.I. OG1E7.RA., Factory Mutual Research Corporation, Norwood, MA, May 1983.

3. Mawhinney, J. R., and R. Solomon, "Water Mist Fire Suppression Systems," *Fire Protection Handbook,* 18th ed., National Fire Protection Association, Quincy, MA, 1997, Section 6, pp. 216–248.

4. Scheffey, J. L., "Foam Extinguishing Agents and Systems," *Fire Protection Handbook,* 18th ed., National Fire Protection Association, Quincy, MA, 1997, Sec. 6, pp. 349–367.

5. Kuchta, J. M., "Investigation of Fire and Explosion Accidents in the Chemical, Mining, and Fuel-Related Industries—A Manual," Bulletin 680, U.S. Bureau of Mines, Washington, D.C., 1985.

6. Zabetakis, M. G., "Flammability Characteristics of Combustible Gases and Vapors," Bulletin 627, U.S. Bureau of Mines, Washington, D.C., 1965.

7. Cote, A. E., ed., *Fire Protection Handbook,* 17th ed., National Fire Protection Association, Quincy, MA, 1991.

8. Taylor, G. M., "Halogenated Agents and Systems," *Fire Protection Handbook,* 18th ed., National Fire Protection Association, Quincy, MA, 1997, Section 6, pp. 281–296.

9. NFPA 12A, *Standard on Halon 1301 Fire Extinguishing Systems,* National Fire Protection Association, Quincy, MA, 1997.

10. Purser, D. A., "Toxicity Assessment of Combustion Products," *The SFPE Handbook of Fire Protection Engineering,* 2nd ed., National Fire Protection Association, Quincy, MA, 1995, Section 2, pp. 85–146.

11. Cogan, D. G., *Stones in a Glass House: CFCs and Ozone Depletion,* Investor Responsibility Research Center, Inc., Washington, D.C, 1988.

12. Huggett, C., "Habitable Atmospheres Which Do Not Support Combustion," *Combustion and Flame,* Vol. 20, 1973, pp. 140–142.

13. DiNenno, P. J., "Direct Halon Replacement Agents and Systems," *Fire Protection Handbook,* 18th ed., National Fire Protection Association, Quincy, MA, 1997, Section 6, pp. 297–330.

14. Ewing, C. T., F. R. Faith, J. T. Hughes, and H. W. Carhart, "Flame Extinguishment Properties of Dry Chemicals," *Fire Technology,* Vol. 25, 1989, pp. 134–149.

15. Iya, K. S., Wollowitz, S., and Kaskan, W. E., "The Mechanism of Flame Inhibition by Sodium Salts," Fifteenth Symposium (International) on Combustion, The Combustion Institute, Pittsburgh, PA, 1975, pp. 329–336.

16. Hertzberg, M., K. L. Cashdollar, and C. P. Lazzara, "The Limits of Flammability of Pulverized Coals and Other Dusts," Eighteenth Symposium (International) on Combustion, The Combustion Institute, Pittsburgh, Pennsylvania, 1981, pp. 717–729.

17. Hague, D. R., "Dry Chemical Agents and Application Systems," *Fire Protection Handbook,* 18th ed., National Fire Protection Association, Quincy, MA, 1997, Section 6, pp. 341–348.

18. Kumar, R. K., Tamm, H., and Harrison, W. C., "Intermediate-Scale Combustion Studies of Hydrogen-Air-Steam Mixtures," NP-2955, Electric Power Research Institute, Palo Alto, CA., 1984.

19. Bajpai, S. N., and J. P. Wagner, "Inerting Characteristics of Halogenated Hydrocarbons," *Industrial & Engineering Chemistry, Product Research & Development,* Vol. 14, 1975, pp. 54–59.

20. Hickey, H. E., "Foam System Calculations," *The SFPE Handbook of Fire Protection Engineering,* 2nd ed., National Fire Protection Association, Quincy, MA, 1995, Section 4, pp. 99–122.

21. Thompson N. J., *Fire Behavior and Sprinklers,* National Fire Protection Association, Quincy, MA, 1964.

22. Cote, A., and P. Bugbee, *Principles of Fire Protection,* National Fire Protection Association, Quincy, MA, 1988.

23. Clark, W. E., *Fire Fighting Principles & Practices,* Dun-Donnelley Publishing Corporation, New York, 1976.

24. Cote, A. E., ed., *Fire Protection Handbook,* 18th ed., National Fire Protection Association, Quincy, MA, 1997.

25. Heskestad, G., "The Role of Water in Suppression of Fire," *Fire and Flammability,* Vol. 11, 1980, pp. 254–259.

26. Rasbash, D. J., "The Extinction of Fire with Plain Water: A Review," *Fire Safety Science: Proceedings of the First International Symposium,* Hemisphere Publishing Company, New York, 1986, pp. 1145–1163.

27. Magee, R. S., and R. D. Reitz, "Extinguishment of Radiation-Augmented Plastics Fires by Water Sprays," *Fifteenth Symposium (International) on Combustion,* The Combustion Institute, Pittsburgh, PA, 1975, pp. 337–347.

28. Wahl, A. M., "Water and Water Additives for Fire Fighting," *Fire Protection Handbook,* 18th ed., National Fire Protection Association, Quincy, MA, 1997, Section 6, pp. 5–14.

29. Tapscott, R. E., "Combustible Metal Extinguishing Agents and Application Techniques," *Fire Protection Handbook,* 18th ed., National Fire Protection Association, Quincy, MA, 1997, Section 6, pp. 401–405.

30. *Fire Protection Guide on Hazardous Materials,* 11th ed., National Fire Protection Association, Quincy, MA, 1994.

CHAPTER 15

Special Fire Situations

In an ordinary fire, the combustible is cellulosic, a synthetic polymer, or a petroleum product, and it was ignited by an external source. It burns in normal air at normal atmospheric pressure. This chapter considers a number of special fire situations deviating from these conditions. Consideration is given to materials capable of spontaneous ignition, to fires involving exothermic materials, and to fires in oxygen-enriched atmospheres, or at pressures other than atmospheric, or in microgravity environments.

SPONTANEOUS IGNITION

Whenever the following four conditions exist, *spontaneous ignition*, also called *spontaneous combustion*, can occur:

1. The material undergoes, however slowly, a heat-generating (exothermic) reaction at its normal temperature with the oxygen of the air.

2. The rate of this reaction increases rapidly with increasing temperature.

3. The size and physical arrangement of the material are such that heat cannot escape readily from its interior.

4. The material, if ignited, is capable of smoldering (it must be porous and must form rigid char).

The condition for spontaneous ignition is charted in Figure 15.1. The slope of the curve depicting the rate of heat generation increases progressively with increasing temperature, as is characteristic of chemical reactions. The two heat-loss curves, for a faster and a slower cooling rate, are approximately linear (constant slope) because, according to *Newton's law of cooling, the rate of cooling is directly proportional to the difference between the temperature of the warm object and the temperature of the surroundings.*

Figure 15.1 shows that, at the initial temperature, the rate of heat generation is finite but the rate of cooling is zero because the object is initially at the same temperature as the surroundings. As a result, the temperature of the object must rise and must continue to rise until the cooling curve intersects the heating curve. The intersection point corresponds to a balance of heating and cooling, and no further temperature increase occurs. The intersection temperature might be only slightly above the initial temperature, and no ignition would result.

Figure 15.1 also shows that if the cooling curve increases too slowly as the object temperature increases, then the heating curve and the cooling curve do not intersect. The object gets hotter and hotter until it ignites.

If the temperature of the surroundings were to be increased, it can be deduced that the heat-generation rate versus temperature would not be affected, but the cooling-rate curve, while retaining its slope, would be shifted to the right. Thus, if the temperature of the surroundings were increased suffi-

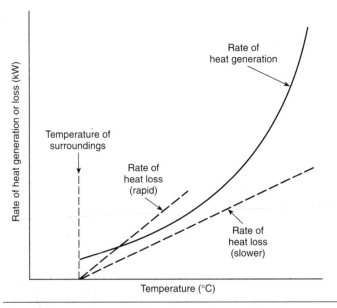

FIGURE 15.1 Relative heat-generation and heat-loss rates, sometimes leading to runaway heating and ignition.

ciently, the upper cooling curve in Figure 15.1 would no longer intersect the heat-generation curve, and ignition would result.

Once combustion commences, and the temperature is at a high level, the rate of heat generation is generally no longer limited by chemical reaction rates, but by the rate at which oxygen can diffuse into the porous object, which does not increase as rapidly with temperature. Meanwhile, the rate of heat loss increases more rapidly with increasing temperature, because very hot objects lose heat by radiation as well as by conduction and convection. Accordingly, the heat-generation and heat-loss curves will intersect at a high temperature for a steadily burning object. *(See Figure 15.2.)*

The foregoing discussion presents a theoretical basis for spontaneous ignition. In reality, however, objects ignite spontaneously only in special cases. Furniture, beds, and newspapers do not burst into flame spontaneously. The reason is that most common solid materials react so slowly with oxygen at normal temperatures that the self-heating, if measurable at all, usually

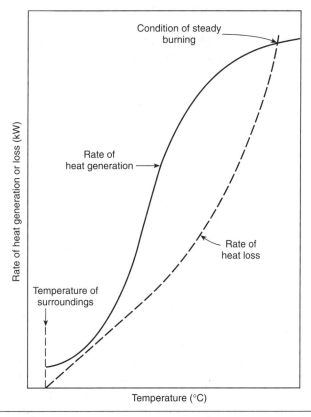

FIGURE 15.2 Balance of heat-generation and heat-loss rates for steady burning.

amounts to no more than one or two degrees. However, there are special cases where spontaneous ignition can occur.

Consider a large pile of coal, sawdust, wood chips, grain, grass clippings, or hay. None of these materials conducts heat well because of the air gaps between the individual particles. Accordingly, even a very slow rate of heat generation in the interior can eventually cause a substantial temperature rise, even though weeks or months could pass before the temperature rose enough to initiate smoldering combustion in the interior. Then, if the pile is disturbed, bringing fresh air in contact with the glowing region in the interior, a sudden transition to flaming combustion can occur. Or, the transition to flaming could occur when the glowing zone progressed from the interior to the surface.

Numerous examples of such fires have been observed. They are especially dangerous in the holds of ships. *The Fire Protection Handbook* [1] (Table A-10) lists approximately 75 substances that are capable of spontaneous heating. Among the most dangerous are rags or other fibrous materials in contact with corn oil, fish oils (e.g., cod liver oil, menhaden oil), linseed oil, perilla oil, pine oil, soybean oil, tung oil, or any unsaturated oil. Such oils are reactive with oxygen at room temperature, and the presence of the rags or fibrous material is needed to provide more surface area for the oil–oxygen reaction, and also to confine the heat and permit the temperature to build up. Saturated oils, on the other hand, such as petroleum-derived oils (e.g., mineral oil, common lubricating oil, or heating oil), do not cause spontaneous heating. Aside from oils, dangerous materials include charcoal and overdried fish meal. Bowes [2] provided extensive details on self-heating behavior of various materials and presented the mathematical theory underlying Figure 15.1.

Bowes [2] described interesting studies of the self-heating and ignition of haystacks. Very little happens unless the moisture content of the hay is greater than about 25 percent, which requires either direct application of water (rain) or continued exposure to very high relative humidity (about 90 percent). It is assumed that the hay was dried before stacking. Under humid conditions, aerobic fungi and bacteria grow in the hay at normal outdoor temperatures, generating heat by oxidation. If the hay is a few meters across, then the temperature will rise to about 75°C in several days or weeks. Above this temperature, the organisms are no longer active. However, a slower chemical oxidation then takes over. It is believed that the oxygen is not reacting with the hay itself at this stage, but with the decomposition products of the hay, which were formed during the biological heating. After several more weeks, the temperature could rise to the point where flaming ignition occurs spontaneously.

Consider Figure 15.1 again. The rate of heat generation within a porous object should be proportional to the volume of the object, while the rate of heat loss should be proportional to its surface area. If d is a characteristic dimension, the volume is proportional to d^3, and the surface area is proportional to d^2. Therefore, the ratio of volume to surface area, or the ratio of heat-generation rate to heat-loss rate, should be proportional to d. Hence, for any material capable of self-heating leading to subsequent ignition, there must be a critical value of d that leads to ignition. For damp haystacks, this critical value appears to be several meters.

Drysdale [3] quoted experiments in which the critical size of cubes of "chemically activated carbon" was measured at different temperatures of the surroundings. Some of the findings follow:

Ambient Temperature	Critical Size	Time to Ignition
60°C	60.1 cm	68 hours
99°C	15.1 cm	14 hours
125°C	5.1 cm	1.3 hours

Given such data (combined with theory), the critical size and time to ignition at lower temperatures, corresponding to normal ambient temperatures, could be calculated. It is evident that the size would be rather large and the time would be rather long.

In general, then, the ability of a material to ignite spontaneously depends on the following properties:

1. The basic nature of the material
2. Its porosity
3. Its size
4. Its moisture content
5. The temperature of its surroundings

The time required could be as long as 1 month. The theory of spontaneous ignition is understood well; however, the biological or chemical rate of heat generation at normal temperatures is unknown for most materials.

EXOTHERMIC MATERIALS

An *exothermic material* is defined as either a pure substance or a mixture of substances that can undergo chemical reactions (liberating heat) without requiring oxygen from the air. If such materials are involved in a fire, they usually will cause easier ignition, faster fire growth, higher flame temperatures, and more difficult extinguishment. In some cases they will permit propagating combustion in the complete absence of oxygen. In many cases they will explode.

There are thousands of such materials to be found among industrial chemicals. Different strategies for fighting fires are appropriate for different materials, depending on quantities involved, likelihood of explosion, toxicity, solubility in water, compatibility with extinguishing agents, and so on. Guidelines for dealing with many exothermic materials can be found in the *Fire Protection Guide to Hazardous Materials* [4] or in Sax and Lewis [5]. Another option is the Chemical Transportation Emergency Center (CHEMTREC), available by phone (1-800-424-9300) day or night, including Sundays, for information relating to hazardous materials emergencies. The most common exothermic materials in various categories will be mentioned in this section.

Energy from Decomposition

Certain chemical compounds are thermodynamically unstable and can decompose into smaller molecules while releasing energy in the form of heat. Examples include the following:

Ozone: $O_3 \rightarrow {}^3/_2\,O_2$
Nitrous oxide: $N_2O \rightarrow N_2 + {}^1/_2\,O_2$
Acetylene: $C_2H_2 \rightarrow 2\,C + H_2$
Cyanogen: $C_2N_2 \rightarrow 2\,C + N_2$
Hydrogen peroxide: $H_2O_2 \rightarrow H_2O + {}^1/_2\,O_2$
Hydrazine: $N_2H_4 \rightarrow N_2 + 2\,H_2$
Propyne: $CH_3CCH \rightarrow 3\,C + 2\,H_2$
Ethylene oxide: $C_2H_4O \rightarrow CO + CH_4$, or $CO + H_2 + {}^1/_2\,C_2H_4$
Lead azide: $Pb(N_3)_2 \rightarrow Pb + 3\,N_2$
Diborane: $B_2H_6 \rightarrow 2\,B + 3\,H_2$
Methyl hydrazine: $N_2H_3CH_3 \rightarrow N_2 + 3\,H_2 + C$, or $2\,NH_3 + C$

Note that, in some cases, the products of reaction depend on the reaction conditions.

Energy from Polymerization

When organic monomers combine with one another to form polymers, energy is released. In controlled polymerization, this energy is removed as it is released, by the circulation of a coolant. In uncontrolled polymerization, excessive temperatures are produced. (See Chapter 10, "Fire Characteristics: Solid Combustibles," for examples of polymeric reactions.)

Energy from Intramolecular Oxidation-Reduction

In some cases, a given molecule can contain a portion with oxidizing power, such as a nitro group (—NO_2) or a peroxide group (—O—O—), and a portion that can be oxidized, which might involve carbon and hydrogen atoms. Such molecules often are used as explosives or rocket propellants. Examples include the following:

Glyceryl trinitrate $\rightarrow$ $^3/_2$ N_2 + $^5/_2$ H_2O + 3 CO_2 + $^1/_2$ O_2
(Nitroglycerin)

Trinitrotoluene $\rightarrow$ 6 CO + $^3/_2$ N_2 + C + $^5/_2$ H_2
(TNT)

RDX $\rightarrow$ 3 N_2 + 3 CO_2 + 3 H_2
or
(Cyclonite)

Nitromethane:

$$CH_3NO_2 \rightarrow CO + H_2O + {}^1/_2 H_2 + {}^1/_2 N_2 \quad \text{or}$$
$$CO_2 + {}^3/_2 H_2 + {}^1/_2 N_2$$

Methyl nitrate:

$$CH_3ONO_2 \rightarrow CO_2 + H_2O + {}^1/_2 H_2 + {}^1/_2 N_2 \quad \text{or}$$
$${}^1/_2 CO_2 + {}^1/_2 CO + {}^3/_2 H_2O + {}^1/_2 N_2$$

Ammonium nitrate:

$$NH_4NO_3 \rightarrow N_2 + 2 H_2O + {}^1/_2 O_2$$

Ammonium perchlorate:

$$NH_4ClO_4 \rightarrow {}^1/_2 N_2 + HCl + {}^3/_2 H_2O + {}^5/_4 O_2$$

Cellulose dinitrate (Pyroxylin® or Collodion®):

$$\rightarrow N_2 + 6 CO + 3 H_2O + {}^1/_2 H_2 \quad \text{or}$$
$$N_2 + 3 CO_2 + 3 CO + {}^7/_2 H_2$$

Energy from an Oxidizing Agent in Contact with a Reducing Agent

Some *oxidizing agents* (e.g., ozone, hydrogen peroxide, ammonium perchlorate, and ammonium nitrate) can decompose exothermically by themselves, but other oxidizing agents (e.g., potassium nitrate, sodium chlorate, and fluorine) are not capable of exothermic decomposition. Nevertheless, an oxidizer of either type, in contact with an oxidizable material *(reducing agent)*, can react to liberate large amounts of energy. Almost any organic material or metal can be oxidized. Gunpowder, certain explosives, pyrotechnic compositions, and composite solid propellants for rockets consist of physical mixtures of oxidizers and oxidizable materials. Examples include the following:

Black powder: $KNO_3 + \frac{3}{2}C + \frac{1}{2}S \rightarrow CO_2, CO, N_2, K_2S, K_2SO_4, K_2CO_3, S$

Ammonium nitrate + fuel oil (ANFO explosive): $n\, NH_4NO_3 + \frac{1}{3}(CH_2)_n \rightarrow n[N_2 + \frac{7}{3}H_2O + \frac{1}{3}CO_2]$

Pyrotechnic delay mixture: $BaCrO_4 + B \rightarrow BaO + \frac{1}{2}B_2O_3 + \frac{1}{2}Cr_2O_3$

Solid propellant: $n(NH_4ClO_4 + Al) + 1.4\,(CH_2)_n \rightarrow N_2, H_2, H_2O, HCl, Al_2O_3, CO, CO_2$

Such mixtures are prepared with extreme care and presumably are stored in bunkers or isolated places. A more commonly encountered problem relates to oxidizers themselves that have not been deliberately mixed with anything but might become mixed by accident, such as in the course of a fire. Accordingly, some of the more common classes of oxidizers are given:

1. Nitric acid and its salts (nitrates)
2. Perchloric acid and its salts (perchlorates)
3. Chromic acid and its salts (chromates and dichromates)
4. Permanganic acid and its salts (permanganates)
5. Fluorine, chlorine, bromine, iodine (in order of decreasing reactivity)
6. Inorganic peroxides
7. Sodium perborate
8. Calcium hypochlorite
9. Potassium persulfate
10. Manganese dioxide
11. Chlorate, bromate, and iodate salts

Any of these oxidizers, in contact with common materials, such as paper, cotton, wood, plastics, hydrocarbons, alcohols, vegetable and animal oils and fats, sulfur, or metals, can react exothermically even in the absence of air, although, in some cases, an elevated temperature is needed to initiate reaction.

Special Cases

It should be noted that the class of organic compounds known as *ethers* has the capability of forming peroxides after storage for a month or more. Isopropyl ether ($C_3H_7OC_3H_7$) is said to be more susceptible to peroxide formation than other ethers. [6] Detonation can occur when this peroxide is heated. Many other peroxides, such as the widely used benzoyl peroxide, (C_6H_5)—$(C=O)$—O—O—$(C=O)$—(C_6H_5), can undergo rapid exothermic decomposition.

Certain metals react with water, nitrogen, or other extinguishing agents. (See Chapter 10, "Fire Characteristics: Solid Combustibles," and Chapter 14, "Fire Fighting Procedures.") It is interesting that magnesium powder will react violently with Halon 1301 (CF_3Br), forming MgF_2, $MgBr_2$, and carbon, while releasing 26,700 J/g. This is more energy than is released by burning magnesium in oxygen. Aluminum powder also will react violently with halogenated liquids, which include various common solvents (e.g., trichloroethylene), as well as halons. Aluminum or magnesium will react exothermically with the oxides of certain other metals, such as iron, lead, tin, or copper, in what is known as the thermite reaction, producing, for example, temperatures high enough to melt steel.

Alkaline substances, also known as caustics or bases, will react with any of the numerous organic and inorganic acids, or even with water, releasing substantial amounts of heat, which could cause other nearby materials to ignite. The most common alkalies are sodium hydroxide (lye) and calcium oxide (lime or quicklime). In solution, these alkalies can generate hydrogen when in contact with aluminum or galvanized steel (zinc). The carbides of lithium, sodium, potassium, calcium, or barium will react with water to form acetylene, which is highly flammable.

Controlled exothermic reactions of many types are encountered in chemical processing plants. If such a process goes out of control, overheating and fire can result.

FIRES IN ABNORMAL ENVIRONMENTS

Oxygen-Enriched Atmospheres

Normal air contains 21 percent oxygen by volume. For special purposes, such as in undersea operations, in spacecraft, or in medical chambers, it might be desirable to maintain a higher percentage of oxygen. However, this higher percentage of oxygen greatly increases the fire hazard.

A higher percentage of oxygen implies a lower percentage of the inert gas nitrogen. This situation causes the flame temperature to be higher because the nitrogen, if present, acts to absorb heat. The higher flame temperature has several consequences:

1. The rate of heat transfer from the flame to the surroundings is greater.
2. The flame is less easily quenched by adjacent cold surfaces, so the stand-off distance of the flame from the surface is less, and the flame can spread more rapidly.

3. The process of soot formation in a diffusion flame is enhanced by higher temperature, so the flame becomes sootier and emits more radiation.

Certain solids that will not burn in normal air, unless preheated, will burn in oxygen-enriched air.

Figure 15.3 shows how the rate of horizontal flame spread over cellulose and over thick Plexiglas increases with increasing oxygen percentage. Additional such data are provided by Dorr [7], Carhart [8], and Fernandez-Pello et al. [9]. It must be noted that the rate of flame spread depends on the thickness and orientation of the specimen as well as on the oxygen percentage.

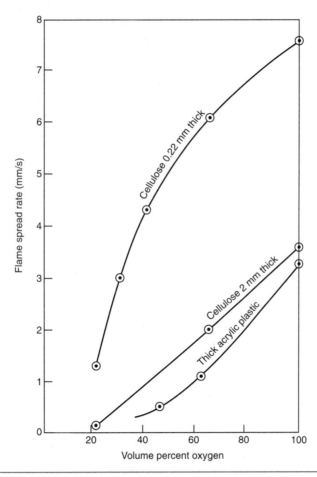

FIGURE 15.3 Effect of oxygen percentage in atmosphere on horizontal flame spread over surfaces. From Lastrina et al [10] and McAlevy and Magee. [11]

In addition to increase in the rate of flame spread, the burning rate per unit area increases at higher oxygen concentrations. Figure 15.4 shows data for rates of burning of small pools of heptane and various plastics in atmospheres with various proportions of oxygen and nitrogen, some richer and some leaner in oxygen than normal air. The sharp increase in burning rate with increasing oxygen percentage is evident; however, somewhat surprisingly, the rates seem to level off above an oxygen level of about 40 percent. Unfortunately, the experiments were not continued to the 100 percent oxygen level. It was found that the flames over these combustibles become progressively sootier as the oxygen content in the atmosphere is increased.

The *limiting oxygen index* (LOI) shows that many plastics and natural materials will not burn in normal air when a pilot flame is applied momentarily to the top of a small isolated sample in the absence of preheating or of external radiative flux. (See Chapter 10, "Fire Characteristics: Solid Combustibles.") Examples of such materials are wool, leather, neoprene rubber, polyvinyl chloride, and Teflon. However, Table 10.3 shows that each of these materials will burn with sufficient oxygen enrichment of the atmosphere. Furthermore, various metals not normally flammable in air will burn in oxygen-enriched air or pure oxygen.

Smoldering combustion occurs more readily in oxygen-enriched air than in normal air. Data by Moussa et al. [13] showed a fourfold increase in the

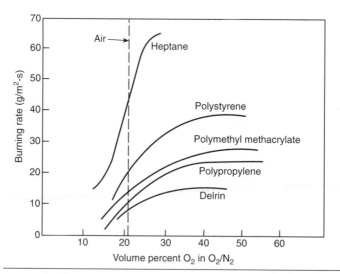

FIGURE 15.4 Effect of oxygen percentage on the burning rate on small, thin, horizontal samples of plastics and heptane. From Tewarson. [12]

smoldering rate of cellulose rods when the oxygen was increased from 21 percent to 96 percent.

For gaseous mixtures, the upper flammability limit is dramatically increased with oxygen enrichment of the atmosphere, but the lower flammability limit is hardly affected because there is an excess of oxygen at this limit anyway. For example, the lower and upper limits for methane in air are 5 percent and 15 percent, respectively, by volume, while the lower and upper limits of methane in oxygen are 5 percent and 61 percent, respectively. In general, the flammability limits are much wider when fuel vapors are mixed with oxygen rather than air. Some additional data are provided by Zabetakis. [14]

Fires in oxygen-enriched atmospheres not only burn hotter and faster but also, not surprisingly, are more difficult to extinguish. However, little data exist to quantify this observation. A series of articles on ignition, burning, and extinguishment of various metals and nonmetals in oxygen-enriched atmospheres, published by American Society for Testing and Materials (ASTM), can be consulted for more information. [15, 16, 18]

Nonatmospheric Pressure Environments

Normally, fires are thought of as occurring at the pressure existing on earth at sea level: 1 atm, or approximately 1 bar (100,000 Pa). However, the pressure at the top of Mt. Everest is only 0.3 atm. The cabin of a commercial aircraft flying at high altitude is normally pressurized to an absolute pressure of 0.68 atm. The pressure 100 m below the surface of the sea is 11 atm, absolute (i.e., 10 atm above the sea-level pressure).

Therefore, in specialized circumstances involving aviation or underwater operations, fires at pressures other than atmospheric might be encountered. Such fires also might occur inside pressurized chambers used in chemical processing or other manufacturing operations, or in hyperbaric chambers in hospitals. If the percentage of oxygen remains at 21 percent, then the flame temperature will be nearly the same for atmospheric and nonatmospheric pressure fires, and the differences in fire behavior are not as profound as when the percentage of oxygen is changed.

One important effect of the change of pressure is the modification of *flammability limits*. Figure 15.5 shows the low-pressure part of the flammability limit diagram for propane–air mixtures. No propane–air mixture is flammable under these test conditions below 0.04 atm absolute pressure. At pressures above about 0.5 atm, flammability limits are virtually independent of pressure.

However, the *fire point* of a liquid, which is essentially the temperature at which the vapor pressure of the liquid equals the lower flammability limit, is highly dependent on pressure. Clearly, if the pressure is at 0.5 atm instead of 1 atm, the vapor pressure of the liquid need be only one-half as great to achieve

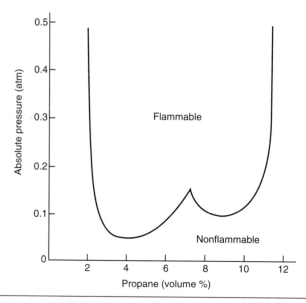

FIGURE 15.5 Low-pressure limit of flame propagation of propane–air mixtures in a 10-cm diameter tube. From DiPiazza et al. [19]

the lower flammable limit. Therefore, flash points and fire points are lower than normal at pressures below atmospheric, and they are higher than normal at pressures above atmospheric. This variation is very important in assessing the flammability in the vapor space of an aircraft fuel tank.

The *minimum ignition energy* that can ignite fuel vapor–air mixtures is lower when the pressure is higher, and higher when the pressure is lower. The minimum spark energy needed for ignition for most vapors is approximately inversely proportional to the square of the absolute pressure. Thus, an extremely low energy spark could ignite a combustible gas mixture at high pressure. At high pressure a higher voltage would be required to cause a spark to jump across a given gap.

The rate of flame spread over solid surfaces increases with increasing pressure. Data from McAlevy and Magee [11] show that the rates of horizontal spread over thick slabs of Plexiglas and polystyrene vary with the 0.82 and 0.76 power of pressure, respectively. Other fire properties, such as ignition temperature, tendency to produce smoke, flame radiation, and toxic gas concentrations, will all vary somewhat with pressure.

Although most atmospheres that contain enough oxygen to be breathable also will support combustion, it is notable that a mixture of 10 percent oxygen and 90 percent nitrogen at 2 atm absolute pressure is easily breathable, but will not support combustion of most materials.

Microgravity Environments

If a fire were to occur in a space station orbiting earth, it would be expected to behave differently than an ordinary fire on earth. The fire products would not be buoyant in orbit because of the virtual absence of a gravitational force, so these products would not rise from the combustion zone to make room for fresh air needed to continue the oxidation. This scenario suggests that fires might not be able to sustain themselves in the absence of gravitation.

Because even a small fire in a space station could have very serious consequences, research efforts are being directed toward this problem. The term microgravity is used instead of zero gravity because of very small residual gravitational effects, present even in orbit, that are caused by high-altitude drag, operation of machinery, and movement of occupants.

In the past few years, a series of combustion experiments have been performed in orbit around the earth, in the NASA Shuttle program. In the orbital vehicles, the gravitational effect is only a few millionths as strong as on earth. Principal findings to date follow:

1. In spite of the absence of buoyancy, flames do not smother themselves but continue to burn by a diffusion process. A wax candle has continued to burn for 45 minutes

2. Flame spread rates across surfaces, and burning rates, are somewhat slower.

3. The flames generally are cooler and sootier than flames in earth's gravity and have different shapes.

4. The burning of some materials in low gravity might result in hot globules of the material drifting in all directions, possibly starting other fires.

5. Under some circumstances it will take less energy to ignite a solid in low gravity because of the reduced convective cooling of the heated surface by the surrounding cold atmosphere.

6. The forced convection imposed by spacecraft ventilation systems can accelerate burning rate and flame spread.

More information can be found in a conference report, "Fourth International Microgravity Combustion Workshop." [20]

The choice of an agent for extinguishment of a fire in a space station must take into account the cleanup problem after the fire. Possibilities of agents with minimal cleanup problems are water sprays and nitrogen. There is, as yet, very little experience with fire fighting in space stations. The atmosphere in a space station is not necessarily 21 percent oxygen and 79 percent nitrogen, and the pressure could be other than 1 atm. Accordingly, the information given earlier in this chapter might apply. (See subsections Oxygen-Enriched Atmospheres and Nonatmospheric Pressure Environments.)

The design of flammability tests for selection of materials to put in a space station is another problem. The difference in smoke movement in a low-gravity environment must be considered when choosing locations for smoke detectors in a space station. If an ultraviolet or infrared flame detector is chosen, the flicker characteristics of the flame will be different, and possibly there will be some spectral differences, compared with flames in earth's gravity.

Problems

1. The tendency of oily rags to ignite spontaneously depends on the type of oil. Why are many vegetable oils and fish oils far more dangerous than mineral oil or lubricating oil?

2. How is it possible for certain liquids and gases, such as ethylene oxide or acetylene, to burn in the complete absence of oxygen?

3. What is special about calcium carbide?

4. Why do marginally combustible materials often burn readily in oxygen-enriched air?

5. What is the effect of oxygen enrichment on flammability limits?

6. How are flammability limits and fire points changed when the pressure is above 1 atm?

7. Is it true that any artificial atmosphere that will support life will also support combustion of common materials?

8. How and why would a fire in an orbiting space station containing air at atmospheric pressure behave differently than a fire on earth?

References

1. Cote, A. E., ed., *Fire Protection Handbook*, 18th ed., National Fire Protection Association, Quincy, MA, 1997.

2. Bowes, P. C., *Self-Heating: Evaluating and Controlling the Hazards*, Elsevier, New York, 1984.

3. Drysdale, D., *An Introduction to Fire Dynamics*, 2nd ed., J. Wiley, New York, 1998.

4. *Fire Protection Guide to Hazardous Materials*, 11th ed., National Fire Protection Association, Quincy, MA, 1994.

5. Sax, N. I., and R. J. Lewis, *Dangerous Properties of Industrial Materials,* 7th ed., 3 Vol., National Fire Protection Association, Quincy, MA, 1996.

6. Bradford, W. J., "Storage and Handling of Chemicals," *Fire Protection Handbook,* 18th ed., National Fire Protection Association, Quincy, MA, 1997, Section 3, pp. 256–266.

7. Dorr, V. A., "Fire Studies in Oxygen-Enriched Atmosphere," *Journal of Fire and Flammability,* Vol. 1, 1970, pp. 91–98.

8. Carhart, H. W., "Inerting and Atmospheres," *Spacecraft Fire Safety,* NASA Conference Publication 2476, 1987, p. 56.

9. Fernández-Pello, A. C., S. R. Ray, and I. Glassman, "Flame Spread in an Opposed Forced Flow: The Effect of Ambient Oxygen Concentration," *Eighteenth Symposium (International) on Combustion,* The Combustion Institute, Pittsburgh, PA, 1981, pp. 579–589.

10. Lastrina, F. A., R. S. Magee, and R. F. McAlevy, "Flame Spread Over Fuel Beds: Solid Phase Energy Considerations," *Thirteenth Symposium (International) on Combustion,* The Combustion Institute, Pittsburgh, PA, 1971, pp. 935–948.

11. McAlevy, R. F., and R. S. Magee, "The Mechanism of Flame Spreading Over the Surface of Igniting Condensed Phase Materials," *Twelfth Symposium (International) on Combustion,* The Combustion Institute, Pittsburgh, PA, 1969, pp. 215–227.

12. Tewarson, A., "Generation of Heat and Chemical Compounds in Fires," *The SFPE Handbook of Fire Protection Engineering,* National Fire Protection Association, Quincy, MA, 1988, Section 1, pp. 179–199.

13. Moussa, N. A., T. Y. Toong, and S. Backer, "An Experimental Investigation of Flame-Spreading Mechanisms Over Textile Materials," *Combustion Science & Technology,* Vol. 8, 1973, pp. 165–175.

14. Zabetakis, M. G., "Flammability Characteristics of Combustible Gases and Vapors," Bulletin 627, U.S. Bureau of Mines, Washington, D.C., 1965.

15. Werley, B., ed., "Flammability and Sensitivity of Materials in Oxygen-Enriched Atmospheres," STP 812, ASTM, W. Conshohocken, PA, 1983.

16. Benning, M. A., ed. "Flammability and Sensitivity of Materials in Oxygen-Enriched Atmospheres," STP 910, ASTM, W. Conshohocken, PA, 1986.

17. Schroll, D. W., ed., "Flammability and Sensitivity of Materials in Oxygen-Enriched Atmospheres," STP 986, ASTM, W. Conshohocken, PA, 1988.

18. Stoltzfus, J. M., and K. McIlroy, "Flammability and Sensitivity of Materials in Oxygen-Enriched Atmospheres," STP IIII, 1991, ASTM, Philadelphia, PA.

19. DiPiazza, J. T., M. Gerstein, and R. C. Weast, "Flammability Limits of Hydrocarbon-Air Mixtures—Reduced Pressures," *Industrial and Engineering Chemistry,* Vol. 43, 1951, pp. 2721–2725.

20. "Fourth International Microgravity Combustion Workshop," NASA Conference Publication 10194, Washington, D.C., May 1997.

Glossary

Absorptivity   The fraction of radiant energy striking a surface that is absorbed, rather than reflected or transmitted into the interior.

Acceleration   The rate of change of velocity.

Acid   A substance that will release hydrogen ions when dissolved in water.

Activation energy   The energy needed to overcome the barrier to chemical reaction when two molecules, or an atom and a molecule, come together.

Addition polymer   A polymer that contains the same atoms originally present in the monomer(s) from which it was formed. *(See Condensation polymer.)*

Alcohol   Any of a number of carbon-hydrogen-oxygen compounds containing one or more hydroxyl groups attached to carbon atoms. Examples are methanol (methyl alcohol), ethanol (ethyl alcohol), isopropanol, ethylene glycol, and glycerol (glycerine). All are flammable liquids.

Alkali   A substance that will react with acids in aqueous solution to form salts, while releasing heat. Examples are lye, lime, and ammonia. Also called bases.

Alkali metal   Any of the elements lithium, sodium, potassium, rubidium, cesium, or francium. They react with water to form alkalis, such as sodium hydroxide ($NaOH$).

Alkaline earths   Any of the elements beryllium, magnesium, calcium, strontium, barium, or radium.

Aqueous   Watery.

Aqueous film-forming foam (AFFF) A foam that forms a spreading aqueous film over the surface of a flammable liquid.

Aromatic hydrocarbon Any of a family of hydrocarbons having a ring structure like that of benzene, diphenyl, naphthalene, anthracene, and so on.

Association reaction The combination of 2 atoms or molecules to form a larger molecule.

Atomic number An integer from 1 to 109, proportional to the positive electric charge on the nucleus of each of the 109 kinds of atoms (elements).

Atomic weight The relative weight of a kind of atom, approximately on a scale of 1 (for hydrogen) to 238 (for uranium). Precisely, the ratio of the weight of the kind of atom to the weight of the C-12 isotope of carbon.

Atom A tiny particle consisting of a nucleus with a positive electric charge, surrounded by negatively charged electrons. Each chemical element consists of characteristic atoms.

Attenuation The progressive reduction in intensity of something, such as a beam of light passing through smoke.

Autoignition Spontaneous ignition, generally of the type resulting from external heating by a hot surface, or by compression.

Beer-Lambert law *(See Bouguer's law.)*

Bernoulli's equation An equation stating the relation between pressure and velocity along a streamline of a fluid.

Bio-assay A procedure for determining toxicity of smoke that requires exposure of test animals, usually rats or mice.

BLEVE Boiling liquid expanded vapor explosion. Can occur with a fire external to a railroad tank car, a tank truck, or a drum of flammable liquid.

Boilover A phenomenon that can occur during a fire over an open tank containing a blend of flammable liquids, such as crude oil; water must be present at the bottom of the tank for boilover to occur.

Bouguer's law A mathematical relationship describing the decrease of intensity of a beam of light as it passes through a semitransparent region, such as smoke. (Also known as the Beer-Lambert law.)

Branching chain reaction A sequence of chemical reactions involving chain carriers (active species; generally free radicals or free atoms) in which chain carriers are produced more rapidly than they are consumed. For example, if A is a reactant, B is a product, and X is a chain carrier, then $A + X \rightarrow B + 2X$ would be a chain-branching step.

Buoyancy The tendency of a less dense region of fluid to rise, relative to surrounding denser fluid, because of gravitation.

Burning rate   The combustion rate of a solid or liquid, expressed either as the rate of mass consumption per unit exposed area, or as the rate of heat released per unit exposed area, in which case it is generally called the heat-release rate.

Burning velocity   The speed with which a laminar flame moves in a direction normal to its surface, relative to the unburned portion of a combustible gas mixture, at constant pressure. Its value depends on mixture composition, temperature, and pressure.

Calorimeter   An apparatus used to measure the rate of thermal energy output of an exothermic process, such as a fire.

Carboxyhemoglobin   A chemical compound resulting from the reaction of hemoglobin with carbon monoxide, after which oxygen can no longer be transported by the hemoglobin.

Ceiling jet   The radially outward flow under a ceiling resulting when a fire plume impinges on a ceiling.

Cellulose   A natural polymer $(C_6H_{10}O_5)_n$, which is a principal constituent of cotton, wood, and paper.

Celsius temperature   A temperature scale on which pure water at sea level freezes at $0°$ and boils at $100°$.

CFAST   A zone-type computer model of a fire in a compartment connected to other compartments.

Chain reaction   The rapid reaction of a free atom or radical with another species, the products of which include 1 or more free atoms or radicals, which can undergo further rapid reactions. If more than 1 free atom or radical is produced, a branching chain reaction results.

Chemical analysis   The experimental procedures by which the identity of the molecules in a sample is determined.

Chemical bond   The attractive force that often binds together two atoms into a combination that is stable at least at room temperature, but which becomes unstable at a sufficiently high temperature.

Chemical compound   A substance involving a tightly bound combination (not a physical mixture) of two or more kinds of chemical elements, in definite proportions.

Chemical equilibrium   A group of molecules at a given temperature and pressure is said to be in a state of chemical equilibrium if no net chemical change would occur, regardless of how long one might wait. The higher the temperature, the shorter the time needed to reach chemical equilibrium.

Chemical kinetics   The study of the factors controlling the rates at which chemical systems change their compositions in their approach to equilibrium.

Chemical thermodynamics   The relationships governing atoms and molecules in a state of chemical equilibrium.

Class A fire   Fire in "ordinary" combustible solids. However, if a plastic readily melts in a fire, it might be Class B rather than Class A.

Class B fire Fire in flammable liquids, gases, and greases.

Class C fire Fire in energized electrical equipment.

Class D fire Fire in combustible metals.

Class I liquid A liquid with flash point below 38°C (100°F).

Class II liquid A liquid with flash point between 38°C and 60°C (100°F and 140°F).

Class III liquid A liquid with flash point at or above 60°C (140°F).

Combustion An exothermic chemical reaction that occurs so rapidly that the heat released causes the temperature of the reaction zone to be many hundreds or even several thousands of degrees higher than the surroundings.

Condensation polymer A polymer that contains fewer atoms than originally present in the monomer(s) from which it was formed, because a small molecule, usually H_2O, separated out each time two units joined together. *(See Addition polymer.)*

Conduction of heat The transfer of heat from the hot to the cold side of a medium by means of energy transfer from molecule to adjacent molecule, or atom to atom.

Conservation of energy The principle specifying that, when one form of energy (kinetic, potential, chemical, thermal, electrical, etc.) is converted into another form of energy, the total energy is unchanged.

Conservation of momentum The principle specifying that the momentum of a moving object, or a moving fluid, defined as the product of the mass and the velocity, remains constant unless an external force, such as mechanical work or friction, acts on the moving object or fluid.

Convective heat Energy that is carried by a hot moving fluid. (Energy can be *radiated* through space, or *conducted* through a solid, as well as *convected* by a fluid.)

Copolymer A polymer made of monomers of two or more different kinds of molecules.

Critical temperature The temperature of a pure substance above which distinct liquid and vapor phases cannot coexist, regardless of the pressure.

Cross-linked polymer A polymer in which the long chains are bonded to one another at intermediate points. Cross-linking reduces flexibility and tendency to melt, and increases the tendency to form char on heating.

Crystal A solid consisting of atoms, molecules, or ions fixed in a regular geometric pattern (cubic, hexagonal, tetrahedral, and so on).

Cyclic hydrocarbon A molecule whose structure includes one or more rings of carbon atoms. Six-membered rings are the most stable. Examples are benzene, toluene, cyclohexane, and naphthalene.

Deep-seated fire A fire burning within a material, as opposed to a surface fire. For example, a fire that has burrowed into a mattress or a carton of documents.

Deflagration A propagating combustion zone moving at subsonic velocity (as contrasted to a detonation).

Density   Usually, mass per unit volume. However, when an extinguishing agent is applied to a surface, the term *density* is used to mean the mass rate of application of agent per unit of surface.

Detonation   A violent type of combustion in a premixed gas or in a liquid or solid explosive that moves at supersonic speed and causes a very large pressure increase.

Diffusion flame   A flame in which the fuel vapors and oxygen diffuse into each other as they burn, for example, a candle flame.

Dissociation reaction   The break-up of a molecule into two fragments.

Dose   In toxic studies involving fire smoke, the mathematical product of concentration of the toxic material and the time of exposure.

Dry chemical   Any of several powders used to extinguish fires.

Dry ice   Solid carbon dioxide.

Electron   A very light particle with a negative electric charge, a number of which surround the nucleus of most atoms. The flow of electricity through a wire or across a vacuum tube, such as a television picture tube, is a flow of electrons.

Elements   The 109 kinds of atoms of which all matter is composed. Each element consists of atoms unique to that element and different from the atoms of all other elements.

Emissivity   The fraction of radiative energy emitted by a given surface area of a hot object, relative to that emitted from an opening of the same area in a furnace at the same temperature. The emissivity is between 0 and 1.

Endothermic reaction   A chemical change that absorbs heat.

Enthalpy change   The energy of the chemical reaction products at a given temperature minus the energy of the reactants at that temperature, for a constant-pressure process. The symbol is ΔH_T. When the value is negative, heat is released, and, when positive, heat is absorbed.

Entrainment   The process in which a slow-moving or stagnant fluid mixes into an adjacent turbulent flow. For example, a rising fire plume entrains surrounding air.

Ether   Any of a number of carbon-hydrogen-oxygen compounds that consist of two carbon-hydrogen groups linked together by an oxygen atom. Examples are diethyl ether, methyl ethyl ether, and diphenyl ether. All are flammable liquids.

Exothermic material   A material that can undergo exothermic reaction without requiring oxygen from the air.

Exothermic reaction   A chemical change that releases heat.

Exponential dependence   Mathematically, the equation $y = a \cdot e^x$ shows that y is exponentially dependent on x. The rate of change of y with respect to x is proportional to y.

Extinction coefficient   The constant in Bouguer's law.

Field model   A type of computer fire model that takes into account variations in either two or three dimensions, dividing the space of interest into a fine grid, as contrasted with a zone model, which is basically one-dimensional.

Fire extinguishment   The cessation of combustion.

Fire plume   A buoyant column of fire gases, usually with flames in the lower portion.

Fire point   The minimum temperature to which a liquid must be heated in a standardized apparatus, so that sustained combustion results when a small pilot flame is applied, as long as the liquid is at normal atmospheric pressure.

Fire retardant   A chemical incorporated to modify combustion properties of materials.

Fire suppression   *Either* extinguishment of a fire as a result of fire fighting activity, *or* reduction of the combustion rate to a small enough value so that the fire is under control. (A large fire, after being suppressed, could continue to smolder for days.)

Firebrand   A flaming or smoldering airborne object emerging from a fire, which can sometimes ignite remote combustibles.

Fire triangle   A diagram showing the three elements needed for a fire to come into existence: (1) a combustible material, (2) oxygen, and (3) heat.

Flame arrestor   A device installed in a pipe or duct to prevent the passage of flame.

Flame-spread rate   The rate of advancement of a flame over a combustible surface, expressed as a velocity.

Flammability limit   The boundary of composition, temperature, or pressure separating flammable and nonflammable gas mixtures containing air or another oxidant.

Flammable   Capable of being ignited.

Flash point   The minimum temperature to which a liquid must be heated, in a standardized apparatus, so that a transient flame moves over the liquid when a small pilot flame is applied.

Flashover   For burning in a room, flashover is the often-sudden transition from local burning to widespread burning of all exposed combustibles. After flashover, flames might be projecting out a door or window.

Fluid mechanics   The branch of physics dealing with prediction of the motion of a fluid being acted upon by pressure differences, gravitational force, or other external forces.

Fourier's law of heat conduction   An equation stating that heat conduction per unit area is directly proportional to the temperature gradient.

Free atom   An atom that would be stable when combined with another atom of the same type, but which at the moment is unattached, either because it has just been formed by a chemical reaction, or because it is at very low pressure or frozen in an inert matrix.

Free radical   A molecular fragment with unsatisfied chemical valences, which is unstable in the sense that it will very rapidly combine with other molecules of the same kind or of different kinds, unless it is frozen, or at extremely low pressure (upper atmosphere).

Gas   A state of matter in which the molecules, which are moving rapidly, are separate from one another except when undergoing collisions. *(See definitions for Liquid and Solid, the other states of matter.)*

Glowing combustion   A term applied to smoldering combustion when a glow (radiation from a hot surface) is visible.

Grashof number   A dimensionless number expressing the ratio between buoyant force and viscous force acting on a fluid in motion.

Gross heat of combustion   The value of the heat of combustion when the water formed is condensed to liquid form. (Also called higher heat of combustion.)

Haber's rule   A given toxic effect is produced by a given mathematical product of concentration and time. Haber's rule is only valid under certain limited conditions. *(See Dose.)*

Halogen   Any of the elements fluorine, chlorine, bromine, iodine, or astatine.

Halons   Various synthetic chemicals, the molecules of which consist of combinations of carbon, fluorine, chlorine, and bromine.

Heat capacity   The energy that must be added to a unit mass of a substance in order to raise its temperature by 1°C (as long as no phase change occurs). Also called thermal capacity. *(See Specific heat.)*

Heat of combustion at constant pressure ($-\Delta H_{298}$)   The energy released when 1 mole of a combustible reacts completely with oxygen at atmospheric pressure and 298 K to form combustion products at 298 K.

Heat of condensation   The energy released when a unit mass of vapor condenses to a liquid.

Heat of fusion   The energy absorbed when a unit mass of a solid melts.

Heat of gasification (of a solid)   The energy absorbed by a solid when it vaporizes or decomposes to form gases.

Heat of solidification   The energy released when a unit mass of a liquid solidifies.

Heat of sublimation   The energy absorbed when a unit mass of a solid gasifies directly, without forming a liquid and without chemical change.

Heat of vaporization   The energy absorbed when a unit mass of a liquid vaporizes.

Heat-release rate   The rate at which thermal energy is released by a fire, often expressed as kilowatts per unit of burning area. It consists of a combination of convective, radiative, and conductive heat. It can be calculated from the mass burning rate if the heat of combustion and combustion efficiency are known.

Heat transfer   The branch of physics dealing with the calculation of the rate at which thermal energy (heat) moves from a hotter to a cooler region.

Hemoglobin A molecule in the blood that carries oxygen from the lungs to cells throughout the body. A typical formula is $[C_{738}H_{1166}O_{208}N_{203}S_2Fe]_4$.

Hydrocarbon Any molecule containing only two kinds of atoms: hydrogen and carbon. Petroleum products are primarily hydrocarbons.

Hydroxyl A free radical, —OH.

Ignition The onset of combustion. *(See Spontaneous ignition, Autoignition, Piloted ignition, and Thermal ignition.)*

Inflammable Not a permissible word, because it introduces confusion as to whether "flammable" or "nonflammable" is meant.

Internal energy change The energy of the chemical reaction products at a given temperature minus the energy of the reactants at that temperature, for a constant-volume process. The symbol is ΔE_T. When the value is negative, heat is released, and, when positive, heat is absorbed.

Intumescent coating A protective chemical coating, which, when heated, internally generates gases and expands, resulting in a thermally insulating crust that contains cavities.

Ion An atom or molecule that has an excess or deficiency of electrons, and therefore has a negative or positive electric charge.

Isomers A pair (or more) of molecules each consisting of the same number and kinds of atoms, but bound together in different ways. The physical and chemical properties of a pair (or trio) of isomers can be quite different from one another.

Isotopes Different forms of the same chemical element, which have the same number of protons but different numbers of neutrons in the nucleus. There are only very slight differences in the physical and chemical properties of isotopes.

Kelvin temperature An absolute temperature scale. Zero K is the lowest possible temperature; 273 K corresponds to 0°Celsius or 32°Fahrenheit.

Kinetic energy The energy associated with the motion of an object or a fluid.

Laminar flow Nonturbulent flow of a fluid. Also called streamline flow.

Le Chatelier's Principle If the pressure acting on a system in equilibrium is increased, the composition of the system will tend to change in the direction corresponding to an increase in density.

Lean mixture A fuel–air mixture containing more air than required to fully oxidize the fuel molecules (as contrasted with a rich mixture or a stoichiometric mixture).

Linear polymer A polymer consisting of long chains of monomer units without any branching of the chains and contrasting with a cross-linked polymer.

Liquid A state of matter in which the molecules are in contact with one another but can move freely among one another. *(See Gas and Solid, the other states of matter.)*

LNG Liquefied natural gas (primarily methane).

LOI The limiting oxygen index, a characteristic of solid combustibles measured in a standard apparatus in which the O_2/N_2 ratio of the atmosphere is varied, to provide a measure of relative flammability. Also called oxygen index (OI).

LPG Liquefied petroleum gas (primarily propane and butane).

Mass The inertial property of an object, which is proportional to the force needed to accelerate the object in the absence of resistance.

Mass optical density A normalized value of the optical density of a smoke cloud, which is intended to be independent of the measuring apparatus.

Methyl A free radical, $—CH_3$. If the free valence is connected to a chlorine atom, for example, then this forms methyl chloride.

Minimum ignition energy The energy, in millijoules, of the weakest spark that is capable of igniting a gas mixture. The minimum ignition energy depends on the composition, temperature, and pressure of the mixture.

Mole A quantity of a chemical substance equal to its molecular weight in grams.

Molecular weight The sum of the atomic weights of the atoms in a molecule of a given kind.

Molecule Two or more atoms tightly bound together by chemical bonds. A polymer molecule can consist of as many as 100,000 atoms.

Momentum The tendency of a moving body to keep moving in the same direction. Momentum is equal to the product of mass and velocity; for a fluid, it is equal to the product of density and velocity.

Monitor A foam cannon capable of projecting foam, sometimes as far as 100 m.

Monomer A small molecule that can combine with other molecules of the same kind (or different kinds) to form very large molecules, called polymers.

Narcotic effect The effect of producing drowsiness and ultimately unconsciousness. Chemical substances in smoke, when inhaled, can enter the bloodstream and interfere with the oxygen supply to the brain, causing narcosis and possibly death.

Net heat of combustion The value of the heat of combustion when the water formed remains as vapor. (Also called lower heat of combustion.)

Neutron A particle with nearly the same mass as the proton, but electrically neutral; it is part of the nucleus of all atoms except the most common isotope of hydrogen.

Newton's law of cooling The rate of cooling of a hot object in air is proportional to the difference in temperature of the object and the air. (This law ignores radiative cooling.)

Nonflaming combustion Combustion without gaseous flame; also called glowing combustion or smoldering. A cigarette burns this way.

Nonflammable Not capable of being ignited.

Nucleus The positively charged particle at the center of each atom, containing more than 99.9 percent of the mass of the atom. It is surrounded by light, negatively charged, electrons.

Nusselt number   A dimensionless number proportional to the ratio of the convective heat transfer coefficient to the thermal conductivity of a fluid. It also turns out to be the ratio of the characteristic dimension of an object immersed in a flow to the boundary layer thickness.

Oligomer   A molecule consisting of a fairly small number of monomer units that are chemically bonded together.

Optical density   A number quantifying the fraction of a beam of light that is unable to pass through a given smoke cloud.

Optically thick radiator   A semitransparent, hot radiating region (such as flame or smoke), for which the intensity of the emerging radiation is independent of the thickness of the region.

Optically thin radiator   A semitransparent, hot radiating region (such as flame or smoke), for which the intensity of the emerging radiation is proportional to the thickness of the region.

Oxidizing agent   A chemical substance that can react with hydrogen or with metals. Examples are oxygen, ozone, nitrate salts, perchlorate salts, and halogens.

Oxygen index (OI)   *(See LOI.)*

PCBs   Polychlorinated biphenyls, a type of liquid used in electric transformers and capacitors. Manufacture is now banned in the United States, but older equipment containing them is still in use.

Perfect gas law   The mathematical relationship between the density of a gas and its pressure, temperature, and molecular weight.

Phase   A physical state of matter: gas, liquid, or solid.

Phase diagram   A temperature versus pressure plot, showing regions where a substance exists as gas, liquid, or solid, and interfaces between these regions.

Phenyl   The grouping C_6H_5—, with a free valence. If the free valence is attached to chlorine, for example, then it is phenyl chloride. The 6 carbon atoms are arranged in a ring, like benzene (C_6H_6).

Piloted ignition   Ignition of a solid or liquid that is caused by a very small flame held near the surface as the temperature of the solid or liquid is increased by a heat source other than the small flame.

Plastic   A synthetic solid consisting primarily of a polymer or blend of polymers of high molecular weight. Generally, the plastic will also contain other materials, such as plasticizers (for flexibility), fillers (to modify mechanical properties or reduce costs), coloring agents, additives to impart resistance to fungi or ultraviolet light, and fire retardants.

Polymer   A large molecule consisting of very many units, called monomers, chemically bound together. There are many natural as well as synthetic polymers.

Potential energy   The energy contained in an object or fluid because of its position in a gravitational field; e.g., its height above sea level. A compressible fluid may also have potential energy because of its pressure.

Prandtl number   A dimensionless number proportional to the ratio of the viscosity to the thermal conductivity, divided by the heat capacity of a fluid. It is near unity for gases.

Premixed flame   A flame burning in a fuel–oxidant mixture that existed before combustion.

Propagating flame   A flame that is moving to an adjacent region containing combustible and oxygen but not yet ignited.

Proton   The positively charged particle that is the nucleus of the most common isotope of hydrogen.

Pyrolyze   To decompose into other molecules when heated. The pyrolysis products often include gases. (Pyrolysis is the process of pyrolyzing.)

Radiation   The transfer of heat in the form of electromagnetic energy, moving with the velocity of light through a vacuum or a transparent material. Radiation is emitted by all objects above the absolute zero of temperature.

Reaction   A chemical process in which atoms and/or molecules combine with each other, or dissociate, to form other chemical species.

Reducing agent   A chemical substance that can react with oxygen, or with a compound containing oxygen, or with a halogen. Examples are hydrogen and the alkali metals.

Reradiation   When a high radiation flux impinges on and is absorbed by a surface, another radiative flux (often smaller) radiates outward from the surface. This is reradiation, its magnitude depending on the temperature and radiative properties of the surface.

Reynolds number   A dimensionless number expressing the ratio between the inertial force and the viscous force acting on a fluid element.

Rich mixture   A fuel–air mixture containing insufficient oxygen to fully oxidize all fuel molecules present to carbon dioxide and water vapor.

Saturated organic molecule   Any carbon compound possessing no double or triple bonds, but only single bonds, between carbon atoms. Such molecules cannot be hydrogenated.

SI units   An internationally accepted system of measurement units, presented in Chapter 1.

Smoke (from fire)   The mixture of tiny particles and gases produced by a fire. The particles consist mainly of soot, aerosol mist, or both.

Smoke-point height   The height of the shortest laminar diffusion flame that will just release black smoke from its tip, for a given combustible gas or vapor.

Smoldering combustion   Combustion of a solid in the absence of a flame. Generally, smoke is produced.

Solid   A state of matter in which the molecules are mostly or completely locked into position relative to one another, as contrasted with a liquid. A solid can be either crystalline and hard, such as salt, or amorphous and soft, such as butter.

Soot   Tiny particles consisting mostly of carbon, often formed in diffusion flames and in very rich premixed flames.

Specific gravity   The ratio of the density of a substance to the density of water at a specified temperature. For gases, the reference substance is sometimes taken to be dry air at a specified temperature and pressure.

Specific heat   The dimensionless ratio of heat capacity of a substance to that of water at a specified temperature. *(See Heat capacity.)*

Spontaneous combustion   Spontaneous ignition caused by self-heating, usually due to slow oxidation.

Spontaneous ignition   Ignition that occurs as a result of progressive heating, as contrasted with instantaneous ignition caused by exposure to a spark or a flame. The spontaneous ignition can result from self-heating caused by slow oxidation, or from an external heat source. *(See Autoignition and Thermal ignition.)*

Stack effect   The difference in pressure between the top and the bottom of a tall column of warm gas, relative to the difference in pressure between top and bottom of an adjacent column of cold gas. If there are connections between the two columns at top and bottom, flow would be induced.

Stefan–Boltzmann law   An equation that specifies the intensity of radiation emitted by any object, in terms of its absolute temperature (K).

Stoichiometric reaction   Chemical A is said to undergo a stoichiometric reaction with chemical B when the proportions of A and B are such that there is no excess of either A or B remaining after the reaction.

Stoichiometry   The procedure for calculating the combining proportions (by mass or volume) of reactants and products of a chemical reaction, on the basis of chemical formulas and atomic weights. The underlying principles are the conservation of atomic species and the conservation of mass.

Sublimation   The evaporation of molecules from a solid to form a gas in the absence of a liquid.

Surface tension   The elasticlike force in the surface of a liquid, tending to minimize the surface area, causing drops to form. Expressed as newtons per meter or dynes per centimeter. (There are 100,000 dynes per newton.)

Synergism   The joint action of agents, which, when acting together, increase the effectiveness of each other.

Thermal conductivity   A property of any solid or fluid, it is the proportionality constant between the rate of heat conduction and the temperature gradient.

Thermal ignition   Spontaneous ignition that occurs as the result of progressive heating by an external heat source.

Thermite reaction   A highly exothermic reaction between a metal and the oxide of another metal; for example, the reaction of molten or powdered magnesium with iron oxide to form magnesium oxide and molten iron.

Triple point   The unique temperature and pressure at which all three phases (gas, liquid, and solid) of a pure substance can coexist.

Turbulent flow   A random or fluctuating state of motion of portions of a fluid.

Unsaturated organic molecule   Any carbon compound containing double or triple bonds between carbon atoms. Such molecules can be hydrogenated and converted to saturated compounds.

Valence   An integer from 1 to 7, characteristic of an atom; it describes the number of chemical bonds the atom can form with other atoms. Some atoms are able to have any of several valences.

Vapor   A gas that is readily condensible to a liquid.

Vapor pressure   The pressure of the vapor over a liquid so that the rate of evaporation is equal to the rate of condensation. The vapor pressure increases sharply with increasing temperature, up to the critical temperature.

Viscosity   A property of a fluid; it is the proportionality constant between an existing velocity gradient and the shearing force needed to produce this gradient.

Weight   The force acting on an object due to a gravitational field.

Wetting agent   A chemical that, when added to water, reduces its surface tension and improves its ability to penetrate crevices.

Yield   The mass of a given product formed in a chemical reaction per unit mass of starting material. The ratio of the actual yield to the theoretical yield calculated by stoichiometry is often expressed as a percentage.

Zone model   A type of computer fire model based on the assumption that the space of interest can be divided into a few zones; within each all properties are uniform. For example, a compartment is assumed to contain a hot layer and a cold layer, with horizontal interface.

Answers to Problems

Chapter 1

1. $198.6°F = 37°C = 310$ K.
   $5°F = 2.78°C = 2.78$ K.
2. 12 m $= 39.3$ ft ; 90 km/h $= 55.9$ mi/h.
3. 60 psi $= 414$ kPa ; 300 U.S. gal/min $= 1135.5$ L/min.
4. $97°F$.
5. 12.21 mm/min.
6. 9553 cal/g ; $17,167$ Btu/lb.

Chapter 2

1. There are 109 officially recognized elements. A given element may possess several isotopes that have slightly different properties.
2. All chemical compounds are molecules. Some molecules, however, are not compounds of two or more elements but are diatomic (or triatomic) elements, like oxygen (or ozone).
3. In general, valences are satisfied in stable compounds, whereas compounds with unsatisfied valences are generally unstable.

Chapter 3

1. Gas molecules touch each other only momentarily, during collisions. In liquids and solids, molecules are in continuous contact with neighboring molecules.

However, molecules in solids, although they may vibrate and rotate, are fixed in space relative to their neighbors, whereas liquid molecules may move about as well as vibrate and rotate.

2. 4.18 percent by volume.

3. 109,200 J.

4. Heat must be removed.

5. In a physical change, say from a gas to a liquid or a solid, the molecules involved are still the same after the change. In a chemical change, the original molecules no longer exist, and new molecules or uncombined elements may form.

6. In general, an increase of temperature sharply increases the rate of chemical change.

7. The gas expands 61 percent.

8. 2.75 g.

9. See the Glossary.

10. $C_6H_6 + 7.5\ O_2 \rightarrow 6\ CO_2 + 3\ H_2O$.

11. 21 percent by volume; 23.3 percent by weight. An oxygen molecule is 32/28 times as heavy as a nitrogen molecule, but, at a given temperature and pressure, the same number of oxygen molecules or nitrogen molecules is present in a given volume of oxygen or nitrogen gas. Furthermore, in a mixture of two gases, each gas exerts its own partial pressure, independent of the other gas, and the sum of the two partial pressures is the total pressure.

12. See the Glossary. Free atoms and free radicals are relevant to fires because they exist in flames and participate in chain reactions, which control the rate of the combustion.

Chapter 4

1. 19.6 m/s.

2. 576 J.

3. 19.6 m.

4. 29,400 pascals above atmospheric pressure, or 29,400 N/m^2 above atmospheric pressure.

5. 7.66 m/s.

6. Turbulence affects the flow pattern; it affects friction with adjacent surfaces; it affects mixing within a fluid; and it affects heat transfer through a fluid.

Chapter 5

1. 46.6 W.

2. At 20°C, air has a density of 1.21 kg/m^3, a viscosity of 0.0000183 $N\text{-}s/m^2$ (from Table 4.1), which is the same as 0.0000183 $kg\text{-}m/s^2$, and thermal conductivity of 0.024 $W/m\text{-}°C$, from Table 5.1, after interpolation. The Reynolds number $dv\rho/\mu =$ (0.01)(5)(1.21)/(0.0000183) = 3306. Taking the Prandtl number as 0.7, the Nusselt number Nu, by Eq. (5.3), is

$$\text{Nu} = 0.68(3306)^{0.466}(0.7)^{0.33} = 26.4$$

Then, $h = 26.4(k/d) = 26.4(0.024)/(0.01) = 63.4 \text{ W}/\text{m}^2\text{-s}$, and the heat flux is, by Eq. (5.2),

$$hA\Delta T = 63.4(3.1416)(0.01)(1)(35) = 69.6 \text{ W}$$

3. 6980 W.
4. Soot particles are the primary emitter. Carbon dioxide gas and water vapor also emit infrared radiation.

Chapter 7

1. Rusting, a slow oxidation, is not combustion. A hydrazine decomposition flame does not involve oxidation.
2. Smoldering can occur in porous materials, which form a carbonaceous char when heated.
3. A fire can be started by a spark, a match, a cigarette, welding, lightning, cooking, space heating, electric appliance or wiring malfunction, friction, or spontaneous self-heating.
4. A portion of the fire-generated heat is transferred to a not-yet-burning portion of the combustible, which increases in temperature until it generates flammable vapors.
5. A fire may be extinguished by (a) cooling; (b) cutting off the air supply; or (c) suppression by chemicals that interfere with combustion chain reactions.

Chapter 8

1. These flames are blue. Air is aspirated into and mixes with the gas feeding the burner, so the flames are premixed.
2. Turbulence increases the rate of combustion of a diffusion flame and increases the rate of heat transfer from the combustion products to any nearby surface.
3. With a spark, hydrogen is much easier to ignite than hexane. With a hot surface, the reverse is true.
4. Never use the word *inflammable*. Use *flammable* or *nonflammable*, as appropriate.
5. Too great an excess of either fuel gas or air in the mixture causes the flame temperature to be too low for combustion (lower than about 1200°C).
6. See the Glossary. A flame may move faster than its burning velocity because of (a) thermal expansion of hot gases behind the flame; (b) inclination of the flame relative to the flow vector into the flame; or (c) turbulence.
7. Yellowness of flames indicates presence of soot particles, characteristic of diffusion flames of carbon-containing combustibles. A yellow flame is a much stronger emitter of radiant heat than a premixed (blue) flame. Smoke is often generated by yellow flames.
8. See the Glossary.

9. See the Glossary. Smoke-point height provides a possible means for classifying combustibles in regard to flame radiant output and smoke generation.

10. See Table 8.5.

11. Flame radiation affects (a) rate of burning of a solid or liquid; (b) rate of spread of a flame; and (c) thermal hazard to exposed persons.

12. (a) Hydrogen flames are almost invisible. (b) They are ignited easily by a weak spark. (c) Hydrogen–air mixtures have very high burning velocity. (d) Hydrogen–air mixtures have very wide flammability limits. (e) Hydrogen–air mixtures, when confined, may detonate.

13. See the Glossary.

Chapter 9

1. If the vapor concentration needed to reach the lower flammability limit is known, then the fire point may be calculated as that temperature at which the vapor pressure is sufficient to produce this vapor concentration.

2. A liquid will burn at ambient temperatures below its flash point if it is on a wick, if it is in the form of a spray or foam, or if it is exposed to the ignition source for long enough so that the liquid becomes preheated to its fire point.

3. Linear burning rate is mass burning rate per unit area divided by liquid density. See Figure 9.3 for dependence on pool size.

4. See Figure 9.6.

5. Boilover may occur for a liquid with a wide distillation range in an open tank, with water at the bottom of the tank.

6. A BLEVE can occur when the following conditions are met: (a) a fire burns under or next to a closed tank containing a flammable liquid; and (b) the venting relief valve is undersized or not present.

Chapter 10

1. Radiative heat loss from the surface will extinguish a single log with cold surroundings.

2. Spontaneous ignition results from oxidative self-heating or from intense radiative heating from a distant source. Piloted ignition results from direct application of a pilot flame or a spark to the surface.

3. According to Figure 10.2, about 18 kW/m^2 is needed. This is about 23 times the intensity of sunlight.

4. Formation of a char layer slows the burning rate.

5. With upward spread, a substantial part of the flame is close to the not-yet-ignited part of the surface. Yes, a hundred-fold factor has been observed.

6. The thinner the material, the more rapidly it can be heated to the ignition condition.

7. From Table 10.2, it is seen that 73 percent of the heat arriving at the surface is radiative and 27 percent is convective.

8. From 5 percent to 25 percent.

9. They all contain substantial proportions of cellulose.

10. Wool contains peptide linkages that contain nitrogen. Partial combustion can lead to formation of highly toxic hydrogen cyanide.

11. Linear polymers melt, whereas cross-linked polymers tend to char.

12. See the Glossary.

13. The oxygen index test measures the minimum oxygen–nitrogen ratio for which a vertical pencillike sample can continue burning at the top. The results do not correlate consistently with other flammability tests.

14. Retardants may (a) promote char formation; (b) decompose, absorbing heat; (c) release gases, which slow the gaseous combustion; or (d) form a barrier. A combination of effects is common.

15. They may increase smoke, cause corrosion, cause toxicity, reduce the strength of wood, interfere with glue or paint.

16. The calorimeter measures heat release rate versus time, giving peak intensity and duration.

17. See the Glossary.

18. When the boiling point of the metal is higher than the flame temperature, surface burning occurs.

19. Two reasons: (a) molten metal may be thrown about, as water turns into steam; (b) hydrogen may be generated.

Chapter 11

1. Black smoke is primarily carbon, produced in sooty flames. White or light gray smoke generally is a fog or aerosol, consisting of droplets of organic liquids, which may originate by decomposition or partial combustion of the original combustible. If inorganic materials are present in the combustible, then they or their oxides may form a light-colored smoke.

2. The smoke particles scatter light.

3. It permits seeing through smoke, helping to locate the seat of the fire, as well as facilitating rescue.

4. It is different for ionization detectors and light-scattering detectors. See the section Smoke Detectors in Chapter 11.

5. A light-scattering detector can be more sensitive for smoldering fires.

6. The chief narcotic gases are carbon monoxide and hydrogen cyanide. The most common irritant gases are hydrogen chloride and acrolein.

7. See the Glossary.

8. Breathing pure oxygen counteracts the effects of carbon monoxide and hydrogen cyanide, as well as the effect of having breathed air deficient in oxygen, because of high carbon dioxide content.

9. The results are highly dependent on the test conditions. Further, there are physiological differences between test animals and humans.

10. From Figure 11.4, it is seen that about 6000 ppm of carbon monoxide (CO) will cause fatality in rats in 30 minutes. Assume that the combustible is polyethylene

[$(CH_2)_n$], the most common plastic. Assume, as a probable worst case, that when it burns, about $1/3$ of the carbon is converted to CO and the other $2/3$ to CO_2. Then, if we have $(6,000/1,000,000)(40) = 0.24$ m^3 of CO in the 40 m^3 room, and it is uniformly mixed, we have fatal conditions for rats and probably humans.

Now, 0.24 m^3 of CO corresponds to $(0.24)(1200$ g/m$^3)$, or 290 g of CO. How much polyethylene has to burn, under the preceding assumptions, to produce this much CO? The equation is

$$CH_2 + \left(\frac{4}{3}\right) O_2 \rightarrow \left(\frac{1}{3}\right) CO + \left(\frac{2}{3}\right) CO_2 + H_2O$$

Then, we see that $12 + 2 = 14$ g of CH_2 will yield $(12 + 16)/3 = 9.3$ g of CO.

Hence, $(14/9.3)(290) = 437$ g of polyethylene, according to these assumptions, can produce a lethal concentration of CO in a 40-m^3 room. (In English units, this is less than a pound of polyethylene burning in a 1400-ft^3 room.)

11. A PCB is a transformer coolant. A transformer might be present in an industrial building or in the basement of an office building or school.

12. Any chlorine-containing substance may decompose in a fire to form hydrogen chloride.

Chapter 12

1. By Eq. (12.1), the flame height is 1.56 m.

2. Use Eq. (12.2), and assume that the plume surface is a cylinder 1 m in diameter and 1.56 m high. Then, the entrainment comes out to be 1.15 kg/s.

3. The plume, upon impinging on the ceiling, makes a 90 degree turn and spreads out radially under the ceiling, forming a ceiling jet.

4. Assume a fire compartment with an open door. If the rate of burning is sufficient, the flow of combustion products out the doorway is so great as to limit the entry of air into the compartment, so that the rate of burning in this choked condition is governed by the size of the door opening and becomes independent of the properties of the combustible.

5. If a tall building is air-conditioned in summer, the density of the air in the building (prior to any fire) is greater than the density of the outside air. The resulting pressure distribution causes air to leak out of the building near the bottom and leak into the building near the top. The reverse is true if the building is heated in winter. This flow will influence the movement of smoke from any fire.

Chapter 13

1. Field models can predict two-dimensional or three-dimensional variations of temperature and velocity near a fire. Zone models are essentially one-dimensional and predict only average values of temperature and velocity in various zones, or average velocity of flow from one zone to another. Field models require a very powerful computer and a skilled programmer, while zone models are much simpler to use and require only an ordinary personal computer.

2. A computer model might be used for the fire-safe design of a building, especially a building with unusual interior configuration or unusual contents. Another use might be to deduce what happened in a destructive fire, for example, in an arson investigation.

3. The fire heat release rate of the initiating item may be measured with a large calorimeter. Many such measurements have already been made, and results are tabulated in the *SFPE Handbook of Fire Protection Engineering*. Once the fire has grown to the post-flashover stage, the subsequent rate of combustion is governed by the sizes of the ventilation openings and may be calculated.

4. A fire in the open may burn faster than the same type of fire in a compartment because of the essentially infinite oxygen supply in the former case. On the other hand, for a fire burning under a ceiling, with the flames mushrooming under the ceiling, there will usually be enhanced radiative feedback to the combustible and a corresponding increase in combustion rate, at least until the fire grows to a size corresponding to a choked condition.

5. A field model is needed for complex geometries such as these.

6. The architect could be given the freedom to design the building with whatever fire safety features he wanted, with the requirement that the resulting design be tested with a computer model to assure a specified safety level. This could provide major savings in building construction costs but is unlikely to be adopted until highly trustworthy computer models are available.

Chapter 14

1. Water can be modified by (a) lowering its freezing point; (b) improving its penetration into porous materials, with wetting agents; (c) thickening it, to promote adherence to surfaces, and to prevent a stream of water from a hose from breaking up, and therefore project further; (d) adding an agent to reduce friction during flow through a hose or pipe; and (e) converting the water to a foam.

2. Large drops are better for penetrating a fire plume. Small drops are better for cooling a fire plume. This suggests use of a bimodal distribution.

3. An AFFF type of foam is best for a gasoline spill fire. If alcohol is present, an alcohol-resistant type of AFFF foam is needed.

4. High-expansion foam is ideal for flooding a confined space, such as the hold of a ship.

5. Yes, if a foam cannon is available and wind does not interfere.

6. The surface tension of the combustible liquid must exceed the surface tension of the foam solution by an amount in excess of the interfacial tension between the two liquids.

7. Reignition is a threat if there is a deep-seated smoldering fire in porous material.

8. A carbon dioxide cylinder contains a liquid, with a little gas on top. Upon discharge, a mixture of liquid drops and gas emerge; the drops undergo evaporative cooling and change into solid carbon dioxide particles, within milliseconds.

9. Water is better than an inert gas, because (a) water cannot asphyxiate, but inert gas can; (b) reignition is less likely; (c) inert gas is impractical for a large fire outdoors or in a large compartment; (d) water can be projected much farther than an

inert gas; and (e) water is much more widely available. However, water can damage some commodities, while inert gases do not. Further, water can cause electric short circuits.

10. Carbon tetrachloride is quite toxic.

11. Because it readily melts, polyethylene should be considered to be in Class B.

12. Potassium bicarbonate is at least four times as effective, on a weight basis, as limestone.

13. Direct application of water can dilute a water-soluble flammable liquid, such as alcohol or acetone, until it is no longer flammable. Foam should be used on a water-insoluble liquid, such as gasoline. Only special foams can be used on alcohol. Water spray can extinguish a fire of a hydrocarbon with a high fire point, by cooling, but unless the spray is fine, burning droplets will be projected when water drops penetrate below the hot flammable liquid and turn to steam.

14. If possible, shut off the gas supply. If not, attack the base of the flame with an agent, projecting the agent in the same direction as the gas flow. Or, inert the compartment. If outdoors, explosives might be used to blow off the flame.

15. Water boils at an ideal temperature (above ambient, but well below decomposition temperatures of most solid combustibles) and has a very high heat of vaporization.

16. Shock hazard is avoided if there is sufficient stand-off distance and absence of salt in water.

17. Calcium carbide, lithium hydride, sodium peroxide, and any toxic chemical that could cause ground water pollution.

Chapter 15

1. Vegetable oils and fish oils are often unsaturated and can react with oxygen at room temperature, generating heat.

2. Certain compounds, including ethylene oxide, acetylene, and hydrazine, decompose exothermically, with substantial release of energy, causing a major rise of temperature.

3. Calcium carbide reacts readily with water to form acetylene gas.

4. Flame temperatures are higher in oxygen-enriched air.

5. Increased oxygen in the atmosphere causes the upper limit of flammability to increase but has very little effect on the lower limit.

6. Flammability limits of gases are essentially unaffected by pressure increases above atmospheric; the fire point of a liquid is increased.

7. At elevated pressure, an atmosphere with reduced percentage of oxygen can support life but not combustion.

8. A fire in an orbiting space station might burn more slowly because of lack of buoyancy, unless a forced ventilation system supplied oxygen to the fire. The flames would be sootier. The flicker characteristics of the flame would be different, which would affect a detector sensitive to flicker. Globules of burning liquid or solid might be ejected and might drift substantial distances.

Index

oxygen-rich, 252–255
Atmospheric pressure, 255–256
Atom(s), 11–14, 262
    free, 17, 266
    mass and size of, 14–15
    nucleus of, 11, 269
    radioactive, 14
    stability of, 14
Atomic number, 11, 13, 262
Atomic weight, 12–14, 262
Atria, 192
Attenuation, 262
Autoignition, 262. *See also* Spontaneous ignition
Automatic sprinkler system, 211, 212, 235
Aviation, fires in, 255–256

B
Bakelite, 142
Bar, 6
Barrier, fire extinguishment with, 81
Bases, 252
Beer-Lambert law, 161, 262
Benzene
    halons and, 226
    heat of combustion of, 41
    inert gases and, 219
    laminar smoke-point height for, 102
    smoke particles from, 162
    structure of, 19
Benzoyl peroxide, 252
Bernoulli's equation, 52–53, 262
Bimolecular exchange reactions, 32
Bio-assay, 168, 262
Blackbody, 66
Blow-off, 233
Boiling liquid expanded vapor explosion (BLEVE), 112, 113, 262
Boiling points (BP), of metals, 153, 154
Boilover, 120–121, 262
Boron, 154
Boron trifluoride, as extinguishing agent, 239
Bouguer's law, 161, 264
Boundary layer, 56
BP (boiling points), of metals, 153, 154
Branching chain reaction, 97, 262
Breathing rate, 74, 171
British thermal units (Btu), 6
Bromine
    as extinguishing agent, 82, 208
    in halogenated agents, 228, 229
Bromochlorodifluoromethane (halon 1211)
    as extinguishing agent, 224–228
    as suppressant gas, 233
Bromotrifluoromethane, 228

Buildings
    smoke movement in, 192–193
    tall, 192–193
Buoyancy, 262
Buoyancy-dominated diffusion flame, 184, 185
Buoyant plume, 56, 184, 185
Burning rate(s), 263
    of liquid pools, 116–118, 119
    in oxygen-rich atmosphere, 254
    of solids, 132–134
Burning velocity, 93–94, 95, 263
1,3-Butadiene
    combustion energy radiated from, 104
    laminar smoke-point height for, 102
Butadiene, heat of gasification of, 132
n-Butane
    isomer of, 22
    laminar smoke-point height for, 102
Butane isomers, 21, 22
n-Butanol, flame-spread rate of, 118–120

C
Calcium, 154
Calcium carbide, extinguishment of, 129
Calcium chloride, as additive to water, 209
Calcium oxide, 252
Calorimeters, 148–150, 263
Carbon
    combustion of, 36–37, 123, 153
    heat of combustion of, 41
Carbon compounds, gasification of, 123
Carbon dioxide
    and breathing rate, 74, 171
    critical temperature for, 220
    as extinguishing agent, 208, 218–223
    flammability limits with addition of, 227, 228
    liquid, 220
    phase diagram of, 220, 221
    in smoke, 171
    structure of, 19
    as suppressant gas, 233
    triple point of, 220
Carbon disulfide
    flammability limits of, 93
    halons and, 225, 226
    inert gases and, 219
    spark ignition energy of, 89
    thermal ignition temperature of, 90
Carbon monoxide
    heat of combustion of, 41
    oxidation or combustion of, 39–40
    in smoke, 167–168, 169–171

Carbon tetrachloride, as extinguishing agent, 208, 224
Carboxyhemoglobin, 167–168, 169–170, 263
Carboxymethyl cellulose (CMC), 210
Catalysis, 179
Caustics, 252
Ceiling, fire plume under, 187–189
Ceiling jet, 187–189, 192, 263
Cells, 58
Cellulose, 135–137, 142, 263
    heat of combustion of, 41
    oxygen index for, 145
    smoke particles from, 164
Celsius scale, 6, 263
Centimeter, 4
CFAST (consolidated model of fire growth and smoke transport), 202–203, 263
Chain reactions, 42, 263
Chair
    cigarette ignition test of, 148, 149
    composition of, 147–148
Charring, 125–126
Chemical(s)
    hazardous, 240
    industrial, 247–248
Chemical analysis, 263
Chemical bonds, 17–20, 263
Chemical change, 32–34
    combining proportions of, 34–38
    energetics of, 39–41
Chemical compounds, 15–16, 263
Chemical elements, 11–14, 265
Chemical equilibrium, 42–44, 263
Chemical kinetics, 42–44, 263
Chemical thermodynamics, 42, 263
Chemical Transportation Emergency Center (CHEMTREC), 248
Chemistry, of fires, 71–75
Chlorine, 12, 140
Chlorobromomethane, as extinguishing agent, 224
Chloroform, as extinguishing agent, 208
Chlorotetrafluoroethane, as extinguishing agent, 229
Choked condition, 191–192
Cigarette ignition test, 148, 149
Class A fire, 231, 263
Class B fire, 231, 264
Class C fire, 231, 264
Class D fire, 232, 264
Class I liquid, 264
Class II liquid, 264
Class III liquid, 264
CMC (carboxymethyl cellulose), 210
Coal mine, 232

Coating, intumescent, 146, 268
Coherent jet, 210
Combining proportions, of chemical reactions, 34–38
Combustible materials, 71–72
Combustion, 77–82
    chemical mechanisms of, 96–102
    defined, 77–78, 264
    flaming, 78–79
    glowing, 267
    heat of
        at constant pressure, 40, 267
        gross (higher), 40, 267
        net, 40–41, 269
    nonflaming, 78–79, 269
    origination of, 79–80
    oxygen and, 77–78
    smoldering, 271
    spontaneous, 124, 245–247, 272
    spread of, 80
    termination of, 80–82
Combustion products, 159–180
    acrolein, 172
    carbon dioxide, 171
    carbon monoxide, 169–171
    damage to materials and equipment, 178–180
    dilution of, 176–178
    hydrogen chloride, 172
    hydrogen cyanide, 171–172
    measurement of particulate content of, 160–162
    mixture of gases and particulates in, 173–174
    oxygen deficiency due to, 173
    smoke, 159–167
    toxicity of, 167–178
    from various materials
        quantities of, 162–164
        toxicity of, 174–176
Combustion reactions, 34–38
Composite materials, 147–152
Computer models, 195–204
    CFAST, 202–203
    field, 196–197, 198, 199
    limitations of, 203–204
    types of, 195–196
    users of, 196
    zone, 197–203
Concrete
    heat capacity of, 31
    thermal conductivity of, 64
Condensation, 27
    heat of, 31, 267

Condensation polymers, 135, 138, 142–144, 264
Conduction, 62, 102
    of heat, 264
Conservation
    of energy, 49, 264
    of momentum, 264
Consolidated model of fire growth and smoke transport (CFAST), 202–203, 263
Constant pressure, 39–40
    heat of combustion at, 267
Convection, 102
Convective heat, 264
Convective heat transfer, 62, 64–65
Conversion factors, 7–8
Coolant, fire extinguishment with, 81
Cooling, Newton's law of, 244, 269
Cooling curve, 244–245
Copolymers, 140, 264
Copper
    heat capacity of, 31
    thermal conductivity of, 64
Copper powder, as extinguishing agent, 239
Coronene, 101
Corridors, 192
Corrosion, 179–180
Corrugated paper, heat of gasification of, 132
Cotton, 134
    fire retardants for, 147
Critical temperature, 264
    for carbon dioxide, 220
Cross-linked polymers, 135, 139–140, 264
Crude oil, 120–121
Crystals, 20–21, 28, 264
Cyanide salts, 129
Cyclic hydrocarbon, 264
Cyclic molecules, 98
Cyclopropane, 21, 22
    isomer of, 22

D
Dacron (polyethylene terephthalate), 143
Decimeter, 4
Decomposition, 33
    energy from, 248
Deep-seated fire, 225, 264
Deep tank fire, 235–236
Deflagration, 94–96, 264
Delrin. See Polyoxymethylene (Delrin)
Density, 265
Density units, 4–5
Detergent, as additive to water, 209
Detonation, 94–96, 265
Diatomic molecules, 15

Dibromotetrafluoroethane (halon 2402), as extinguishing agent, 224, 225
m-Dichlorobenzene, 22
o-Dichlorobenzene, 22
p-Dichlorobenzene, 22
Dichlorobenzene isomers, 21–23
Diesel fuel, flash point of, 115
Diffusion flames, 86, 265
    buoyancy-dominated, 184, 185
    heights of, 183–185
    laminar, 183–184
    momentum-dominated, 184–185
    soot in, 96–102
    stabilization of, 87, 88
    turbulent, 184–185
Dikes, 236
Dilution, extinguishment by, 81
Dilution ratio, for toxic smoke, 176–178
Dimers, 138
Dimethyl ether, 21, 22
    isomer of, 22
Dissociation reaction, 32, 265
Distillation range, 120
Doorway
    in computer modeling, 200–202
    open, 189–190, 192
    pressure gradients at, 190, 191
Dose, 169, 265
Double bonds, 18, 19
Drop-size distribution, 211, 212
Dry chemical agents, 265
    for extinguishment, 229–231
Dry ice, 220–221, 265
Duct fires, 233

E
Elastomers, 144
Electrical energy, 6
Electric charge, 11
Electric shock hazard, 238
Electron(s), 11, 265
Electronic equipment
    damage to, 179–180
    extinguishment of fire involving, 222, 230
Elements, 11–14, 265
Emissivity, 66, 265
Endothermic reactions, 34, 265
Energetics, of chemical change, 39–41
Energy
    activation, 261
    conservation of, 49, 264
    from decomposition, 248
    from intramolecular oxidation-reduction, 248–250
    kinetic, 49–50, 51–53, 268

for deep tank fire, 235–236
electric shock hazard from, 238
extinguishing agents, 207–232
   aqueous foams, 213–218, 219
   dry chemical, 229–231
   fire classes and, 231–232
   halogenated, 223–229
   ideal, 207–208
   inert gases, 218–223
   steam, 220
   water, 207–212
   water mist, 212–213
for gaseous flames, 232–234
for shallow spill fire, 234–235
water damage from, 238
water incompatibility in, 238–240
Fire gases, movement of, 183–193
Fire plume, 266
   buoyancy dominated, 56, 184, 185
   under ceiling, 187–189
   defined, 186
   entrainment rate of, 187
   height of, 183–185
   laminar vs. turbulent, 56
   momentum-dominated, 184–185
   in open, 186–187
   structure of, 186–189
Fire point, 114, 266
   in nonatmospheric pressure environ-
     ments, 255–256
Fire products. *See* Combustion products
Fire products collector, 148, 150
Fire retardant(s), 144–147, 266
   and smoke, 164
Fire retardant (FR) polyurethane, 144
Fire spread, 80, 102
Fire suppression, 266
   methods of, 72–74
Fire triangle, 71, 72, 266
Flame(s)
   burning velocity of, 93–94, 95
   color of, 44, 78, 86
   defined, 78
   diffusion, 86, 265
     heights of, 183–185
   gaseous, 78, 85–87
   laminar, 86–87
   premixed, 85–86, 271
   propagating, 87, 271
   radiation from, 102–105
   stabilization of, 87, 88
   stationary, 87
   turbulent, 86–87
Flame arrestor, 233–234, 266
Flame spread

direction of, 129–131
   in nonatmospheric pressure environ-
     ments, 256
   in oxygen-rich atmosphere, 253
Flame-spread rates, 266
   over liquid surfaces, 118–120
   over solids, 129–131
Flaming combustion, 78
Flammability, 266
   of solids, 129
Flammability limits, 91–93, 266
   in nonatmospheric pressure environ-
     ments, 255, 256
Flammable liquid spray fire, 235–236
Flammable materials, 90–91
Flashover, 150–151, 152, 266
   post-, 191–192
Flash point, 114–116, 266
Flow
   laminar, 55–56
   turbulent, 55–58
Flow patterns, calculation of, 58–59
Fluid(s). *See also* Liquid(s)
   behavior of, 51–59
   density of, 51
   laminar and turbulent flow of, 55–59
   pressure of, 51
   viscosity of, 53–55
Fluid mechanics, 74–75, 266
Fluorine, in halogenated agents, 229
Fluoroprotein foaming agents, 215
Foaming agents, 81, 213–218, 219
   alcohol-type, 216–217
   application of, 218, 219
   applications of, 213–214
   aqueous film-forming, 215–216
   characteristics of, 214
   fluoroprotein, 215
   medium- and high-expansion, 217–218
   protein, 215
Foot-pounds, 6
Force units, 5–6
Forest fires, 210
Formica, 142
Fourier's law, of heat conduction, 63, 266
Free atoms, 17, 266
Free radicals, 17, 18–19, 267
Freezing, 28, 29–30
Freezing-point additives, 209
Friction-reducing additives, 210–211
FR (fire retardant) polyurethane, 144
Fuel oil, flash point of, 115
Fuel-shedding property, 215
Furnishings, 147–152
Fusion, heat of, 30, 267

n-Hexane
  burning rate of, 116–117
  flammability limits of, 93
  flash point of, 115
  halons and, 226
  inert gases and, 219
  spark ignition energy of, 89
High-expansion foaming agents, 217–218
Horsepower, 6
Hose, diameter of, 210–211
Hot layer, 189–191, 200–202
Hydrated alumina, 146, 147
Hydrocarbon(s), 16, 268
  aromatic, 98, 101, 262
  cyclic, 264
Hydrochlorofluorocarbons, as extinguishing
    agents, 229
Hydrogen, 205–206
  atom of, 11, 12, 14–15
  combustion energy radiated from, 104
  combustion of, 34–36
  flammability limits of, 93
  halons and, 225, 226
  heat of combustion of, 41
  inert gases and, 219
  from metal and water, 153–154
  spark ignition energy of, 89
  structure of, 18
  thermal ignition temperature of, 90
Hydrogen bromide, 227, 228
Hydrogen chloride
  in smoke, 172
  structure of, 18
Hydrogen cyanide
  in smoke, 168, 171–172, 174
  water in formation of, 239
Hydrogen fires, 225, 226, 238
Hydrogen fluoride, 227, 228
Hydrogen iodide, 228
Hydrogen-oxygen reaction, 97, 98
Hydrogen peroxide, 15, 17
Hydroxyl radical, 18–19, 268
Hydroxymethyl phosphonium chloride, 147

I
Ice, 29–30
  dry, 220–221, 265
Ignition, 268
  of gases, 87–90
  piloted, 124, 270
  process of, 79–80
  of solids, 123–124
  spontaneous, 124, 243–247, 272
  sustained, 124
  thermal, 272

Ignition energy, minimum, 269
  atmospheric pressure and, 256
  spark, 88–89
Ignition studies, of composite materials, 148,
    149
Ignition temperatures, 89–90
Industrial chemicals, 247–248
Inert gases, for extinguishment, 218–223
Inflammable materials, 90–91, 268
Infrared light, visibility through smoke with,
    165
Inorganic chemicals, 238–239
Intensive properties, 61
Internal energy change, 40, 268
Intramolecular oxidation-reduction, energy
    from, 248–250
Intumescent coatings, 146, 268
Ion, 16, 268
Ionization detectors, 166–167
Iron, 155
Iron oxide, magnesium and, 154
Irritant gases, 74, 172
Isobutane, 21, 22
Isobutylene, laminar smoke-point height for,
    102
Isomers, 21–23, 268
Isopropyl ether, 251
Isotopes, 12, 268

J
Jet aviation fuel, flash point of, 115
Joule, 6

K
Kelvins, 7
Kelvin scale, 6–7, 268
Kerosene, flash point of, 115
Kilogram, 4, 5
Kilometer, 4
Kinetic energy, 49–50, 51–53, 268

L
Laminar flames, 86–87
Laminar flow, 55–56, 268
Laminar smoke-point height, 102
Lean mixture, 268
Leather, oxygen index for, 145
Le Châtelier's principle, 32, 268
Length units, 4
Lexan (polycarbonate), 143
  oxygen index for, 145
Lignin, 135–136
Lime, 252
Limestone, as fire retardant, 146

flammability limits and propagation rates of, 90–96

stabilization of, 87

Pressure units, 6

Propagating flames, 87, 271

Propane, 112

    combustion energy radiated from, 104

    combustion of, 36

    flammability limits of, 93

    halons and, 226

    heat of combustion of, 41

    inert gases and, 219

    laminar smoke-point height for, 102

    smoke particles from, 162

    spark ignition energy of, 89

    thermal ignition temperature of, 90

Propylene, 21, 22

    combustion energy radiated from, 104

    isomer of, 22

    laminar smoke-point height for, 102

    structure of, 19

Protein foaming agents, 215

Proton, 12, 271

psi (pounds per square inch), 6

PVC. *See* Polyvinyl chloride (PVC)

Pyrene, 101

Pyrolysis, 34, 78, 123–124, 271

Q

Quicklime, 252

R

Radiant energy, 63

Radiation, 62–63, 271

    in computer modeling, 202

    from flames, 102–105

    re-, 271

    thermal, 62–63, 65–67

Radiative flux, 124, 126

Radiative heat loss, 124

Radiative heat transfer, 62–63, 65–67

Radiator

    optically thick, 270

    optically thin, 270

Radical, free, 267

Reactions, 32–34, 271

  combining proportions of, 34–38

Reducing agents, 250–251, 271

Reignition, 226, 232, 235

Reradiation, 271

Reynolds number, 54–55, 271

Rich mixture, 271

Rigid body

    acceleration of, 47–48

    gravitation effect on, 48–49

laws governing motions of, 47–50

Rock dust, 232

Rubber foam, oxygen index for, 145

Runaway heating, 244

Runoff, contaminated, 239

S

Salts

    as additives to water, 209, 211

    as extinguishing agents, 239

Saran (polyvinylidene chloride), 141

    oxygen index for, 145

Saturated organic molecule, 271

Scattering effect, of smoke, 165, 166

Self-contained breathing apparatus (SCBA), 218

Self-heating, 243–247

Shallow spill fire, 234–235

Silicon, 154

Silk, 136

Single bonds, 17, 18

Singly ionized, 16

SI units, 3–8, 271

    conversion factors for, 7–8

    energy, 6

    force and pressure, 5–6

    length, area, and volume, 4

    mass and density, 4–5

    power, 6

    temperature, 6–7

    time, 5

Slow oxidative reaction, 77

Smoke, 159–180

    acid mist (aerosol) from, 164

    acrolein in, 172

    carbon dioxide in, 171

    carbon monoxide in, 169–171

    characteristics of, 74

    damage to materials and equipment by, 178–180

    defined, 159–160, 271

    dilution of, 176–178

    eye irritants in, 165

    filling of fire compartment by, 189–192

    fire retardants and, 164

    hydrogen chloride in, 172

    hydrogen cyanide in, 171–172

    measurement of particulate content of, 160–162

    mixture of gases and particulates in, 173–174

    movement in buildings of, 192–193

    opacity of, 161

    optical density of, 161–162

    oxygen deficiency due to, 173

Titanium, 154, 155
Toluene
    flash point of, 115
    laminar smoke-point height for, 102
Toxic chemicals, runoff of, 239
Toxic gases, 74
Toxicity, of fire products, 167–178
1,2,4-Trichlorobenzene, as extinguishing
        agent, 208
Tricresyl phosphate
    as fire retardant, 147
    flash point of, 115
Trifluoromethane, as extinguishing agent, 229
Trimers, 138
Triple point, 273
    of carbon dioxide, 220
Tris phosphate, as fire retardant, 147
Turbulent entrainment, 176–178
Turbulent flames, 86–87
Turbulent flow, 55–58, 273
Turpentine, flash point of, 115

U
Underwater operations, fires in, 255–256
Unsaturated organic molecule, 273
Upholstered furniture, 147–152
Urea-formaldehyde polymers, 142
Urea + potassium bicarbonate, as extinguish-
        ing agent, 231

V
Valence, 13, 18–20, 273
Vapor(s), 92, 275
    flammability limits of, 92–93
Vaporization, heat of, 30, 269
Vapor pressure, 27–28, 113–114, 273
Ventilation, and flashover, 150
Ventilation system, 192
Vinyl, 138
Vinyl bromide, as fire retardant, 147
Vinyl polymers, 138–140
    fire retardants for, 147
Viscosity, 53–55, 273
Viscosity additives, 210
Visibility, through smoke, 165
Volume units, 4

W
Water

additives to, 209–211
chemical reactions with, 74
contaminated, 239
for extinguishment, 207–212, 236–237
freezing point of, 209
heat capacity of, 31
reaction of metals with, 153–154
structure of, 18
surface tension of, 211
thickened, 210
vapor pressure of, 28
viscosity of, 54, 210
Water damage, 238
Water-delivery rate, 212, 236–237
Water droplets, size of, 211, 212
Water incompatibility, 238–240
Water mist, for extinguishment, 212–213
Water vapor, 27, 30, 35–36
    heat capacity of, 31
    as suppressant gas, 233
Watts, 6
Weight, 4–5, 273
Wetting agent, 209, 211, 273
Wick, 116
Wind, 192
Windows, open, 192
Wood
    burning of, 135–137
    charring of, 125–126
    fire retardants for, 146
    gasification of, 135–136
    heat capacity of, 31
    heat of combustion of, 41, 136
    heat of gasification of, 132
    oxygen index for, 145
    smoke particles from, 163, 164
    thermal conductivity of, 64
Wool, 136
    oxygen index for, 145
Wrapping paper, 135

Y
Yield, 273

Z
Zirconium, 154
Zone model, 273
Zone models, 191, 197–203